ADDITION FAST FACTS

Timed Math Facts Practice Tests

This book belongs to:

Printed in the United States of America
ISBN: 978-1-64824-043-0
EDventure Learning LLC
5601 State Route 31 #1296
Clay, NY 13039

www.edventurelearning.com

Email us at hello@edventurelearning.com
Follow us on social media @edventurelearning

Welcome to *Addition Fast Facts!* This workbook is designed to help you master addition by covering all facts from 0 to 20 using 100 timed tests. By practicing regularly, you'll strengthen your addition skills, increase your speed, and build the confidence you need to succeed in math.

How to Use This Workbook:

- Set a Timer: Each test is designed to be completed as quickly as possible. Use a timer to track your speed and challenge yourself to improve.
- Practice Regularly: Consistency is key! Work on one or two tests each day to steadily improve your skills.
- Review Mistakes: After each test, check the answer key in the back, review any errors, and focus on the facts that need extra practice.
- Track Your Progress: At the top of each page is space to record your time and score. In the back of the book, you will also find a progress tracking chart you can use to record your growth over time. Use it to see how far you've come and to set goals for improvement. Remember, progress takes time—keep practicing, and you'll see great results!

Good luck, and have fun mastering your addition facts!

CONTENTS

1

Add 0-5

Time: ________

Score: _____/ 40

① 4 + 5	② 1 + 3	③ 1 + 2	④ 0 + 3	⑤ 1 + 5
⑥ 3 + 3	⑦ 5 + 0	⑧ 2 + 1	⑨ 3 + 5	⑩ 4 + 3
⑪ 5 + 5	⑫ 0 + 5	⑬ 1 + 4	⑭ 1 + 2	⑮ 0 + 2
⑯ 3 + 2	⑰ 5 + 1	⑱ 5 + 3	⑲ 2 + 0	⑳ 2 + 5
(21) 1 + 1	(22) 3 + 4	(23) 3 + 1	(24) 1 + 0	(25) 4 + 2
(26) 0 + 1	(27) 2 + 4	(28) 4 + 4	(29) 4 + 1	(30) 0 + 0
(31) 5 + 2	(32) 2 + 3	(33) 3 + 0	(34) 2 + 2	(35) 2 + 1
(36) 5 + 4	(37) 4 + 2	(38) 2 + 4	(39) 1 + 3	(40) 5 + 0

2

Add 0-5

Time: ________

Score: ______/ 40

(1) 5 + 4 = ___	(2) 0 + 3 = ___	(3) 2 + 0 = ___	(4) 5 + 2 = ___	(5) 5 + 3 = ___
(6) 4 + 0 = ___	(7) 5 + 5 = ___	(8) 2 + 2 = ___	(9) 4 + 5 = ___	(10) 3 + 1 = ___
(11) 1 + 4 = ___	(12) 4 + 4 = ___	(13) 0 + 2 = ___	(14) 3 + 2 = ___	(15) 0 + 1 = ___
(16) 3 + 0 = ___	(17) 3 + 4 = ___	(18) 1 + 1 = ___	(19) 1 + 5 = ___	(20) 5 + 0 = ___
(21) 0 + 4 = ___	(22) 4 + 3 = ___	(23) 4 + 0 = ___	(24) 4 + 1 = ___	(25) 1 + 0 = ___
(26) 3 + 3 = ___	(27) 2 + 5 = ___	(28) 3 + 5 = ___	(29) 5 + 1 = ___	(30) 0 + 0 = ___
(31) 2 + 3 = ___	(32) 0 + 4 = ___	(33) 4 + 2 = ___	(34) 0 + 1 = ___	(35) 4 + 0 = ___
(36) 3 + 4 = ___	(37) 0 + 5 = ___	(38) 0 + 0 = ___	(39) 0 + 4 = ___	(40) 1 + 3 = ___

3

Add 0-5

Time: __________

Score: _____/ 40

(1) 5 + 1	(2) 4 + 0	(3) 2 + 3	(4) 4 + 5	(5) 0 + 4
(6) 0 + 5	(7) 2 + 2	(8) 1 + 1	(9) 5 + 3	(10) 2 + 5
(11) 4 + 3	(12) 3 + 3	(13) 5 + 4	(14) 1 + 4	(15) 3 + 0
(16) 0 + 2	(17) 5 + 2	(18) 4 + 4	(19) 1 + 0	(20) 5 + 5
(21) 3 + 2	(22) 0 + 3	(23) 1 + 3	(24) 1 + 5	(25) 2 + 4
(26) 3 + 5	(27) 4 + 1	(28) 2 + 0	(29) 4 + 0	(30) 5 + 5
(31) 1 + 4	(32) 1 + 1	(33) 2 + 3	(34) 5 + 0	(35) 3 + 4
(36) 5 + 1	(37) 4 + 5	(38) 3 + 2	(39) 0 + 4	(40) 2 + 5

4

Add 0-5

Time: ________

Score: _____/ 40

(1) 0 + 3	(2) 3 + 4	(3) 3 + 5	(4) 0 + 1	(5) 5 + 3
(6) 2 + 0	(7) 5 + 1	(8) 1 + 4	(9) 4 + 2	(10) 0 + 5
(11) 1 + 3	(12) 0 + 0	(13) 1 + 2	(14) 0 + 2	(15) 4 + 4
(16) 3 + 0	(17) 2 + 4	(18) 1 + 0	(19) 3 + 1	(20) 3 + 3
(21) 2 + 2	(22) 2 + 1	(23) 5 + 2	(24) 5 + 4	(25) 1 + 3
(26) 2 + 0	(27) 4 + 2	(28) 3 + 3	(29) 0 + 3	(30) 1 + 2
(31) 2 + 1	(32) 1 + 1	(33) 4 + 3	(34) 5 + 0	(35) 0 + 1
(36) 1 + 0	(37) 1 + 5	(38) 2 + 5	(39) 4 + 4	(40) 3 + 5

5

Add 0-5

Time: ________

Score: _____/ 40

(1) 3 + 0 = ___	(2) 1 + 5 = ___	(3) 2 + 4 = ___	(4) 2 + 3 = ___	(5) 4 + 3 = ___
(6) 5 + 4 = ___	(7) 4 + 1 = ___	(8) 3 + 2 = ___	(9) 3 + 5 = ___	(10) 4 + 5 = ___
(11) 5 + 0 = ___	(12) 5 + 5 = ___	(13) 0 + 2 = ___	(14) 1 + 4 = ___	(15) 5 + 2 = ___
(16) 2 + 2 = ___	(17) 1 + 2 = ___	(18) 2 + 1 = ___	(19) 0 + 0 = ___	(20) 3 + 1 = ___
(21) 2 + 3 = ___	(22) 4 + 2 = ___	(23) 2 + 0 = ___	(24) 5 + 1 = ___	(25) 3 + 3 = ___
(26) 3 + 4 = ___	(27) 4 + 1 = ___	(28) 2 + 2 = ___	(29) 1 + 2 = ___	(30) 4 + 3 = ___
(31) 0 + 5 = ___	(32) 0 + 2 = ___	(33) 5 + 4 = ___	(34) 0 + 0 = ___	(35) 2 + 1 = ___
(36) 0 + 3 = ___	(37) 0 + 1 = ___	(38) 1 + 2 = ___	(39) 2 + 4 = ___	(40) 1 + 3 = ___

6

Add 0-5

Time: ___________

Score: _____/ 40

(1) 5 + 5	(2) 0 + 3	(3) 1 + 5	(4) 4 + 4	(5) 3 + 5
(6) 0 + 5	(7) 5 + 2	(8) 4 + 5	(9) 1 + 0	(10) 3 + 2
(11) 1 + 1	(12) 1 + 4	(13) 1 + 3	(14) 5 + 3	(15) 2 + 5
(16) 4 + 0	(17) 0 + 4	(18) 3 + 3	(19) 2 + 3	(20) 2 + 4
(21) 5 + 1	(22) 3 + 1	(23) 0 + 0	(24) 4 + 0	(25) 1 + 5
(26) 5 + 4	(27) 3 + 0	(28) 2 + 2	(29) 4 + 3	(30) 5 + 5
(31) 3 + 5	(32) 0 + 3	(33) 3 + 4	(34) 0 + 2	(35) 4 + 1
(36) 5 + 2	(37) 0 + 4	(38) 1 + 3	(39) 4 + 0	(40) 4 + 5

7

Add 0-5

Time: ________

Score: _____/ 40

(1) 2 + 3	(2) 4 + 4	(3) 1 + 0	(4) 5 + 0	(5) 1 + 1
(6) 2 + 0	(7) 4 + 2	(8) 5 + 3	(9) 0 + 5	(10) 4 + 1
(11) 2 + 5	(12) 0 + 1	(13) 5 + 0	(14) 1 + 4	(15) 4 + 1
(16) 3 + 3	(17) 0 + 4	(18) 5 + 5	(19) 2 + 4	(20) 2 + 2
(21) 1 + 0	(22) 0 + 5	(23) 3 + 1	(24) 0 + 3	(25) 1 + 4
(26) 1 + 1	(27) 5 + 2	(28) 5 + 0	(29) 0 + 2	(30) 4 + 0
(31) 5 + 1	(32) 3 + 2	(33) 3 + 5	(34) 4 + 1	(35) 3 + 3
(36) 5 + 3	(37) 0 + 4	(38) 0 + 5	(39) 2 + 0	(40) 1 + 4

8

Add 0-5

Time: __________

Score: _____/ 40

(1) $\begin{array}{r} 2 \\ +\ 4 \\ \hline \end{array}$	(2) $\begin{array}{r} 0 \\ +\ 3 \\ \hline \end{array}$	(3) $\begin{array}{r} 3 \\ +\ 3 \\ \hline \end{array}$	(4) $\begin{array}{r} 5 \\ +\ 2 \\ \hline \end{array}$	(5) $\begin{array}{r} 4 \\ +\ 2 \\ \hline \end{array}$
(6) $\begin{array}{r} 2 \\ +\ 4 \\ \hline \end{array}$	(7) $\begin{array}{r} 2 \\ +\ 5 \\ \hline \end{array}$	(8) $\begin{array}{r} 1 \\ +\ 1 \\ \hline \end{array}$	(9) $\begin{array}{r} 1 \\ +\ 4 \\ \hline \end{array}$	(10) $\begin{array}{r} 4 \\ +\ 0 \\ \hline \end{array}$
(11) $\begin{array}{r} 5 \\ +\ 2 \\ \hline \end{array}$	(12) $\begin{array}{r} 4 \\ +\ 2 \\ \hline \end{array}$	(13) $\begin{array}{r} 0 \\ +\ 4 \\ \hline \end{array}$	(14) $\begin{array}{r} 2 \\ +\ 5 \\ \hline \end{array}$	(15) $\begin{array}{r} 2 \\ +\ 4 \\ \hline \end{array}$
(16) $\begin{array}{r} 5 \\ +\ 0 \\ \hline \end{array}$	(17) $\begin{array}{r} 3 \\ +\ 3 \\ \hline \end{array}$	(18) $\begin{array}{r} 1 \\ +\ 1 \\ \hline \end{array}$	(19) $\begin{array}{r} 0 \\ +\ 5 \\ \hline \end{array}$	(20) $\begin{array}{r} 2 \\ +\ 2 \\ \hline \end{array}$
(21) $\begin{array}{r} 5 \\ +\ 0 \\ \hline \end{array}$	(22) $\begin{array}{r} 0 \\ +\ 5 \\ \hline \end{array}$	(23) $\begin{array}{r} 0 \\ +\ 2 \\ \hline \end{array}$	(24) $\begin{array}{r} 5 \\ +\ 1 \\ \hline \end{array}$	(25) $\begin{array}{r} 3 \\ +\ 0 \\ \hline \end{array}$
(26) $\begin{array}{r} 4 \\ +\ 1 \\ \hline \end{array}$	(27) $\begin{array}{r} 2 \\ +\ 0 \\ \hline \end{array}$	(28) $\begin{array}{r} 1 \\ +\ 0 \\ \hline \end{array}$	(29) $\begin{array}{r} 1 \\ +\ 4 \\ \hline \end{array}$	(30) $\begin{array}{r} 4 \\ +\ 1 \\ \hline \end{array}$
(31) $\begin{array}{r} 5 \\ +\ 2 \\ \hline \end{array}$	(32) $\begin{array}{r} 0 \\ +\ 3 \\ \hline \end{array}$	(33) $\begin{array}{r} 3 \\ +\ 1 \\ \hline \end{array}$	(34) $\begin{array}{r} 3 \\ +\ 3 \\ \hline \end{array}$	(35) $\begin{array}{r} 3 \\ +\ 5 \\ \hline \end{array}$
(36) $\begin{array}{r} 1 \\ +\ 1 \\ \hline \end{array}$	(37) $\begin{array}{r} 5 \\ +\ 5 \\ \hline \end{array}$	(38) $\begin{array}{r} 4 \\ +\ 2 \\ \hline \end{array}$	(39) $\begin{array}{r} 1 \\ +\ 4 \\ \hline \end{array}$	(40) $\begin{array}{r} 2 \\ +\ 3 \\ \hline \end{array}$

9

Add 0-5

Time: ________

Score: _____/ 40

(1) 4 + 2	(2) 3 + 3	(3) 4 + 0	(4) 3 + 5	(5) 1 + 1
(6) 3 + 3	(7) 5 + 5	(8) 2 + 4	(9) 4 + 1	(10) 1 + 0
(11) 4 + 2	(12) 1 + 4	(13) 1 + 1	(14) 0 + 2	(15) 2 + 4
(16) 5 + 2	(17) 4 + 1	(18) 1 + 4	(19) 5 + 3	(20) 0 + 3
(21) 2 + 3	(22) 5 + 0	(23) 5 + 1	(24) 4 + 2	(25) 0 + 4
(26) 4 + 1	(27) 3 + 3	(28) 2 + 5	(29) 4 + 0	(30) 3 + 1
(31) 0 + 5	(32) 0 + 4	(33) 5 + 2	(34) 2 + 0	(35) 5 + 0
(36) 3 + 0	(37) 3 + 3	(38) 3 + 5	(39) 2 + 2	(40) 0 + 5

10

Add 0-5

Time: ________

Score: _____/ 40

① 0 + 4	② 4 + 1	③ 4 + 0	④ 1 + 4	⑤ 5 + 0
⑥ 4 + 1	⑦ 5 + 2	⑧ 3 + 1	⑨ 2 + 4	⑩ 5 + 5
⑪ 1 + 0	⑫ 1 + 4	⑬ 4 + 2	⑭ 0 + 5	⑮ 2 + 4
⑯ 2 + 0	⑰ 3 + 5	⑱ 1 + 1	⑲ 2 + 3	⑳ 0 + 3
(21) 1 + 1	(22) 5 + 3	(23) 3 + 3	(24) 4 + 1	(25) 2 + 2
(26) 5 + 1	(27) 0 + 2	(28) 2 + 5	(29) 5 + 2	(30) 3 + 3
(31) 3 + 3	(32) 3 + 0	(33) 0 + 4	(34) 5 + 0	(35) 0 + 5
(36) 4 + 2	(37) 2 + 0	(38) 0 + 2	(39) 5 + 0	(40) 2 + 5

11

Add 0-5

Time: ________

Score: _____/ 40

(1) 3 + 5	(2) 1 + 4	(3) 4 + 2	(4) 1 + 0	(5) 4 + 1
(6) 3 + 1	(7) 2 + 4	(8) 0 + 5	(9) 5 + 0	(10) 5 + 3
(11) 4 + 2	(12) 2 + 4	(13) 0 + 3	(14) 3 + 3	(15) 3 + 0
(16) 5 + 1	(17) 3 + 3	(18) 1 + 1	(19) 4 + 0	(20) 0 + 5
(21) 1 + 4	(22) 0 + 4	(23) 5 + 2	(24) 5 + 5	(25) 3 + 3
(26) 4 + 0	(27) 4 + 1	(28) 1 + 1	(29) 2 + 3	(30) 2 + 5
(31) 1 + 4	(32) 2 + 2	(33) 1 + 1	(34) 4 + 2	(35) 3 + 5
(36) 1 + 5	(37) 4 + 1	(38) 5 + 0	(39) 1 + 4	(40) 3 + 3

12

Add 0-5

Time: ________

Score: _____/ 40

(1) 5 + 0 =	(2) 3 + 5 =	(3) 5 + 2 =	(4) 4 + 0 =	(5) 0 + 3 =
(6) 4 + 2 =	(7) 3 + 1 =	(8) 2 + 3 =	(9) 0 + 2 =	(10) 0 + 4 =
(11) 3 + 3 =	(12) 2 + 4 =	(13) 2 + 5 =	(14) 4 + 2 =	(15) 1 + 1 =
(16) 3 + 0 =	(17) 2 + 2 =	(18) 0 + 5 =	(19) 1 + 4 =	(20) 4 + 0 =
(21) 4 + 1 =	(22) 1 + 4 =	(23) 5 + 2 =	(24) 1 + 0 =	(25) 2 + 0 =
(26) 3 + 3 =	(27) 5 + 5 =	(28) 5 + 0 =	(29) 4 + 1 =	(30) 0 + 4 =
(31) 3 + 2 =	(32) 3 + 3 =	(33) 0 + 2 =	(34) 0 + 5 =	(35) 5 + 0 =
(36) 5 + 3 =	(37) 2 + 5 =	(38) 0 + 3 =	(39) 3 + 3 =	(40) 1 + 1 =

13

Add 0-5

Time: ________

Score: _____/ 40

(1) 4 + 2 = ___	(2) 0 + 3 = ___	(3) 4 + 1 = ___	(4) 1 + 4 = ___	(5) 2 + 4 = ___
(6) 1 + 0 = ___	(7) 4 + 1 = ___	(8) 5 + 5 = ___	(9) 5 + 0 = ___	(10) 1 + 1 = ___
(11) 5 + 2 = ___	(12) 3 + 1 = ___	(13) 4 + 0 = ___	(14) 2 + 2 = ___	(15) 2 + 5 = ___
(16) 3 + 5 = ___	(17) 1 + 4 = ___	(18) 4 + 2 = ___	(19) 2 + 4 = ___	(20) 3 + 3 = ___
(21) 0 + 5 = ___	(22) 0 + 2 = ___	(23) 0 + 4 = ___	(24) 5 + 1 = ___	(25) 3 + 0 = ___
(26) 4 + 1 = ___	(27) 4 + 0 = ___	(28) 2 + 3 = ___	(29) 5 + 0 = ___	(30) 3 + 3 = ___
(31) 5 + 2 = ___	(32) 1 + 1 = ___	(33) 1 + 4 = ___	(34) 4 + 1 = ___	(35) 2 + 2 = ___
(36) 1 + 5 = ___	(37) 0 + 4 = ___	(38) 2 + 0 = ___	(39) 0 + 2 = ___	(40) 2 + 5 = ___

14

Add 0-5

Time: ________

Score: ______/ 40

(1) 0 + 3	(2) 5 + 2	(3) 0 + 4	(4) 5 + 0	(5) 3 + 3
(6) 1 + 1	(7) 0 + 5	(8) 4 + 2	(9) 1 + 0	(10) 3 + 1
(11) 1 + 4	(12) 2 + 4	(13) 3 + 5	(14) 4 + 1	(15) 5 + 3
(16) 4 + 2	(17) 3 + 3	(18) 5 + 0	(19) 2 + 4	(20) 5 + 5
(21) 0 + 2	(22) 1 + 3	(23) 0 + 5	(24) 4 + 0	(25) 3 + 3
(26) 1 + 4	(27) 4 + 2	(28) 3 + 3	(29) 2 + 4	(30) 4 + 2
(31) 2 + 5	(32) 5 + 5	(33) 3 + 5	(34) 0 + 1	(35) 5 + 2
(36) 3 + 0	(37) 2 + 5	(38) 3 + 3	(39) 2 + 3	(40) 2 + 2

15

Add 0-5

Time: ________

Score: _____/ 40

(1) 5 + 1	(2) 3 + 5	(3) 4 + 1	(4) 1 + 1	(5) 4 + 0
(6) 1 + 1	(7) 2 + 0	(8) 0 + 3	(9) 1 + 4	(10) 0 + 4
(11) 5 + 0	(12) 4 + 2	(13) 0 + 5	(14) 4 + 1	(15) 1 + 4
(16) 5 + 0	(17) 2 + 4	(18) 3 + 3	(19) 4 + 1	(20) 5 + 3
(21) 0 + 5	(22) 2 + 3	(23) 5 + 2	(24) 3 + 5	(25) 4 + 0
(26) 0 + 4	(27) 1 + 0	(28) 1 + 4	(29) 1 + 5	(30) 0 + 2
(31) 2 + 2	(32) 4 + 0	(33) 4 + 1	(34) 3 + 0	(35) 0 + 3
(36) 4 + 2	(37) 3 + 3	(38) 2 + 5	(39) 3 + 3	(40) 2 + 4

16

Add 0-5

Time: __________

Score: _____/ 40

① 1 + 4	② 5 + 5	③ 1 + 1	④ 5 + 0	⑤ 3 + 1
⑥ 0 + 5	⑦ 2 + 5	⑧ 5 + 0	⑨ 4 + 1	⑩ 2 + 4
⑪ 1 + 1	⑫ 2 + 0	⑬ 0 + 2	⑭ 0 + 3	⑮ 5 + 2
⑯ 4 + 2	⑰ 0 + 5	⑱ 2 + 2	⑲ 5 + 0	⑳ 2 + 5
(21) 3 + 3	(22) 1 + 3	(23) 3 + 5	(24) 1 + 4	(25) 2 + 4
(26) 0 + 4	(27) 1 + 0	(28) 1 + 1	(29) 1 + 1	(30) 5 + 5
(31) 4 + 2	(32) 4 + 1	(33) 1 + 4	(34) 5 + 1	(35) 2 + 0
(36) 5 + 3	(37) 4 + 1	(38) 2 + 3	(39) 3 + 3	(40) 0 + 5

17

Add 0-5

Time: ________

Score: _____/ 40

(1) $4 + 2$	(2) $4 + 1$	(3) $4 + 0$	(4) $5 + 0$	(5) $0 + 3$
(6) $5 + 2$	(7) $0 + 4$	(8) $3 + 3$	(9) $1 + 1$	(10) $2 + 4$
(11) $0 + 5$	(12) $4 + 2$	(13) $3 + 1$	(14) $3 + 3$	(15) $3 + 0$
(16) $0 + 2$	(17) $1 + 4$	(18) $2 + 5$	(19) $4 + 1$	(20) $1 + 1$
(21) $5 + 0$	(22) $0 + 5$	(23) $1 + 4$	(24) $4 + 1$	(25) $2 + 2$
(26) $2 + 4$	(27) $3 + 3$	(28) $5 + 2$	(29) $3 + 3$	(30) $3 + 5$
(31) $4 + 0$	(32) $5 + 5$	(33) $0 + 3$	(34) $1 + 0$	(35) $1 + 4$
(36) $5 + 0$	(37) $4 + 2$	(38) $2 + 0$	(39) $2 + 5$	(40) $0 + 4$

18

Add 0-5

Time: ________

Score: _____/ 40

① 4 + 0 = ___	② 5 + 1 = ___	③ 3 + 5 = ___	④ 2 + 3 = ___	⑤ 0 + 4 = ___
⑥ 4 + 0 = ___	⑦ 1 + 5 = ___	⑧ 0 + 3 = ___	⑨ 1 + 0 = ___	⑩ 1 + 4 = ___
⑪ 2 + 4 = ___	⑫ 0 + 5 = ___	⑬ 4 + 2 = ___	⑭ 5 + 2 = ___	⑮ 5 + 0 = ___
⑯ 4 + 1 = ___	⑰ 0 + 2 = ___	⑱ 3 + 3 = ___	⑲ 2 + 2 = ___	⑳ 5 + 3 = ___
(21) 2 + 4 = ___	(22) 3 + 0 = ___	(23) 1 + 1 = ___	(24) 1 + 1 = ___	(25) 1 + 4 = ___
(26) 3 + 1 = ___	(27) 2 + 5 = ___	(28) 4 + 1 = ___	(29) 0 + 5 = ___	(30) 5 + 0 = ___
(31) 1 + 4 = ___	(32) 3 + 5 = ___	(33) 0 + 4 = ___	(34) 5 + 2 = ___	(35) 3 + 3 = ___
(36) 4 + 2 = ___	(37) 5 + 5 = ___	(38) 3 + 3 = ___	(39) 3 + 2 = ___	(40) 2 + 0 = ___

19

Add 0-5

Time: ________

Score: _____/ 40

(1) 0 + 5	(2) 2 + 2	(3) 3 + 5	(4) 2 + 0	(5) 1 + 4
(6) 0 + 3	(7) 2 + 4	(8) 4 + 2	(9) 2 + 5	(10) 5 + 0
(11) 5 + 1	(12) 2 + 3	(13) 3 + 3	(14) 4 + 1	(15) 1 + 4
(16) 5 + 2	(17) 3 + 3	(18) 2 + 5	(19) 0 + 2	(20) 3 + 1
(21) 0 + 4	(22) 1 + 1	(23) 0 + 5	(24) 5 + 0	(25) 4 + 1
(26) 1 + 0	(27) 5 + 3	(28) 4 + 0	(29) 2 + 4	(30) 3 + 3
(31) 0 + 4	(32) 1 + 1	(33) 1 + 4	(34) 5 + 5	(35) 3 + 0
(36) 4 + 2	(37) 3 + 5	(38) 2 + 4	(39) 5 + 3	(40) 3 + 3

20

Add 0-5

Time: ________

Score: _____/ 40

(1) $5 + 5$	(2) $2 + 4$	(3) $1 + 0$	(4) $5 + 0$	(5) $0 + 5$
(6) $4 + 0$	(7) $1 + 1$	(8) $3 + 3$	(9) $5 + 2$	(10) $1 + 4$
(11) $0 + 2$	(12) $0 + 3$	(13) $0 + 4$	(14) $4 + 2$	(15) $2 + 4$
(16) $4 + 1$	(17) $2 + 0$	(18) $3 + 0$	(19) $1 + 4$	(20) $5 + 0$
(21) $4 + 0$	(22) $2 + 3$	(23) $4 + 1$	(24) $3 + 3$	(25) $0 + 5$
(26) $2 + 5$	(27) $1 + 1$	(28) $5 + 1$	(29) $3 + 1$	(30) $5 + 2$
(31) $2 + 2$	(32) $1 + 4$	(33) $5 + 5$	(34) $0 + 2$	(35) $5 + 0$
(36) $3 + 2$	(37) $0 + 5$	(38) $0 + 1$	(39) $2 + 2$	(40) $3 + 5$

21

Add 0-10

Time: ________

Score: _____/ 40

(1) $2 + 2$	(2) $1 + 5$	(3) $1 + 3$	(4) $4 + 1$	(5) $6 + 7$
(6) $4 + 4$	(7) $4 + 7$	(8) $9 + 10$	(9) $10 + 6$	(10) $6 + 4$
(11) $0 + 10$	(12) $6 + 9$	(13) $2 + 3$	(14) $10 + 0$	(15) $4 + 0$
(16) $7 + 1$	(17) $9 + 9$	(18) $2 + 9$	(19) $2 + 8$	(20) $0 + 10$
(21) $8 + 7$	(22) $9 + 7$	(23) $3 + 5$	(24) $10 + 0$	(25) $8 + 2$
(26) $7 + 6$	(27) $1 + 9$	(28) $5 + 2$	(29) $9 + 8$	(30) $1 + 4$
(31) $8 + 6$	(32) $5 + 3$	(33) $5 + 5$	(34) $3 + 8$	(35) $10 + 10$
(36) $4 + 1$	(37) $5 + 0$	(38) $4 + 4$	(39) $4 + 8$	(40) $3 + 0$

22

Add 0-10

Time: ________

Score: _____/ 40

① 10 + 7	② 1 + 0	③ 7 + 5	④ 8 + 5	⑤ 0 + 5
⑥ 0 + 2	⑦ 10 + 4	⑧ 3 + 9	⑨ 9 + 5	⑩ 2 + 6
⑪ 6 + 2	⑫ 4 + 6	⑬ 3 + 3	⑭ 5 + 1	⑮ 8 + 2
⑯ 1 + 2	⑰ 0 + 9	⑱ 1 + 10	⑲ 3 + 10	⑳ 2 + 8
(21) 10 + 8	(22) 2 + 4	(23) 8 + 8	(24) 1 + 1	(25) 9 + 4
(26) 2 + 5	(27) 6 + 4	(28) 0 + 1	(29) 9 + 2	(30) 5 + 4
(31) 10 + 1	(32) 8 + 1	(33) 3 + 7	(34) 4 + 2	(35) 6 + 0
(36) 9 + 1	(37) 5 + 8	(38) 6 + 6	(39) 7 + 10	(40) 8 + 10

23

Add 0-10

Time: ________

Score: _____/ 40

(1) 8 + 4	(2) 2 + 10	(3) 8 + 0	(4) 8 + 9	(5) 7 + 3
(6) 5 + 10	(7) 0 + 0	(8) 7 + 0	(9) 6 + 10	(10) 2 + 1
(11) 8 + 3	(12) 4 + 8	(13) 4 + 5	(14) 6 + 8	(15) 4 + 3
(16) 10 + 5	(17) 5 + 6	(18) 3 + 1	(19) 3 + 4	(20) 7 + 7
(21) 0 + 8	(22) 10 + 9	(23) 2 + 0	(24) 9 + 6	(25) 9 + 3
(26) 0 + 6	(27) 6 + 5	(28) 9 + 0	(29) 8 + 0	(30) 7 + 2
(31) 5 + 9	(32) 0 + 3	(33) 7 + 4	(34) 7 + 8	(35) 4 + 3
(36) 4 + 10	(37) 4 + 9	(38) 5 + 7	(39) 7 + 9	(40) 1 + 7

24

Add 0-10

Time: ________

Score: _____/ 40

① 0 + 4	② 7 + 9	③ 8 + 2	④ 5 + 5	⑤ 8 + 0
⑥ 5 + 4	⑦ 6 + 0	⑧ 9 + 1	⑨ 0 + 8	⑩ 7 + 8
⑪ 8 + 4	⑫ 4 + 4	⑬ 5 + 9	⑭ 8 + 9	⑮ 6 + 9
⑯ 0 + 0	⑰ 10 + 7	⑱ 9 + 4	⑲ 7 + 1	⑳ 4 + 5
(21) 2 + 9	(22) 4 + 2	(23) 0 + 2	(24) 4 + 8	(25) 9 + 2
(26) 5 + 7	(27) 5 + 3	(28) 10 + 9	(29) 10 + 10	(30) 3 + 0
(31) 10 + 6	(32) 4 + 1	(33) 2 + 6	(34) 8 + 1	(35) 8 + 6
(36) 8 + 8	(37) 0 + 6	(38) 4 + 4	(39) 7 + 10	(40) 1 + 5

25

Add 0-10

Time: ________

Score: _____/ 40

(1) 6 + 8	(2) 5 + 8	(3) 4 + 6	(4) 10 + 5	(5) 10 + 4
(6) 1 + 8	(7) 7 + 5	(8) 5 + 1	(9) 9 + 10	(10) 7 + 3
(11) 4 + 7	(12) 4 + 9	(13) 8 + 1	(14) 7 + 4	(15) 9 + 5
(16) 6 + 1	(17) 10 + 1	(18) 8 + 3	(19) 6 + 4	(20) 1 + 6
(21) 0 + 4	(22) 0 + 7	(23) 6 + 5	(24) 4 + 0	(25) 1 + 10
(26) 1 + 1	(27) 7 + 7	(28) 0 + 1	(29) 2 + 5	(30) 7 + 0
(31) 0 + 4	(32) 2 + 0	(33) 4 + 6	(34) 6 + 3	(35) 0 + 7
(36) 6 + 10	(37) 4 + 8	(38) 6 + 6	(39) 8 + 7	(40) 9 + 3

26

Add 0-10

Time: __________

Score: _____/ 40

(1) 8 + 10	(2) 9 + 8	(3) 0 + 9	(4) 10 + 3	(5) 0 + 8
(6) 1 + 9	(7) 1 + 7	(8) 0 + 3	(9) 6 + 4	(10) 7 + 6
(11) 10 + 0	(12) 5 + 0	(13) 5 + 6	(14) 8 + 5	(15) 3 + 3
(16) 9 + 9	(17) 4 + 3	(18) 9 + 0	(19) 0 + 5	(20) 3 + 1
(21) 6 + 7	(22) 8 + 0	(23) 3 + 9	(24) 7 + 2	(25) 9 + 6
(26) 0 + 10	(27) 5 + 2	(28) 3 + 5	(29) 10 + 0	(30) 9 + 7
(31) 3 + 8	(32) 2 + 8	(33) 7 + 6	(34) 1 + 0	(35) 8 + 2
(36) 2 + 1	(37) 3 + 7	(38) 1 + 3	(39) 0 + 6	(40) 6 + 2

27

Add 0-10

Time: ____________

Score: ______/ 40

(1) 5 + 10	(2) 10 + 8	(3) 0 + 7	(4) 7 + 4	(5) 7 + 3
(6) 7 + 7	(7) 0 + 1	(8) 9 + 10	(9) 5 + 6	(10) 4 + 6
(11) 6 + 2	(12) 10 + 9	(13) 0 + 2	(14) 7 + 0	(15) 9 + 4
(16) 8 + 10	(17) 3 + 5	(18) 7 + 9	(19) 9 + 7	(20) 0 + 10
(21) 10 + 5	(22) 7 + 1	(23) 7 + 8	(24) 9 + 3	(25) 10 + 0
(26) 5 + 2	(27) 3 + 9	(28) 9 + 0	(29) 1 + 5	(30) 1 + 1
(31) 1 + 10	(32) 6 + 7	(33) 7 + 10	(34) 3 + 3	(35) 5 + 9
(36) 6 + 8	(37) 2 + 8	(38) 6 + 0	(39) 4 + 8	(40) 8 + 4

28

Add 0-10

Time: __________

Score: _____/ 40

(1) 8 + 4	(2) 2 + 5	(3) 0 + 9	(4) 10 + 0	(5) 2 + 2
(6) 8 + 8	(7) 1 + 8	(8) 8 + 10	(9) 1 + 3	(10) 10 + 1
(11) 8 + 3	(12) 2 + 0	(13) 4 + 7	(14) 8 + 1	(15) 1 + 7
(16) 0 + 3	(17) 7 + 6	(18) 8 + 0	(19) 2 + 6	(20) 1 + 9
(21) 10 + 10	(22) 8 + 7	(23) 4 + 8	(24) 0 + 8	(25) 5 + 0
(26) 2 + 1	(27) 5 + 5	(28) 1 + 2	(29) 1 + 10	(30) 9 + 1
(31) 6 + 9	(32) 8 + 0	(33) 10 + 2	(34) 4 + 9	(35) 1 + 0
(36) 6 + 3	(37) 0 + 10	(38) 10 + 0	(39) 6 + 7	(40) 0 + 5

29

Add 0-10

Time: __________

Score: _____/ 40

(1) 3 + 0	(2) 5 + 8	(3) 8 + 2	(4) 8 + 1	(5) 9 + 8
(6) 2 + 8	(7) 5 + 1	(8) 7 + 5	(9) 6 + 5	(10) 4 + 10
(11) 6 + 4	(12) 8 + 9	(13) 1 + 8	(14) 4 + 6	(15) 5 + 4
(16) 6 + 6	(17) 0 + 4	(18) 0 + 0	(19) 3 + 8	(20) 5 + 7
(21) 4 + 1	(22) 5 + 3	(23) 4 + 4	(24) 3 + 1	(25) 9 + 2
(26) 9 + 5	(27) 3 + 7	(28) 4 + 5	(29) 2 + 3	(30) 5 + 10
(31) 10 + 6	(32) 2 + 9	(33) 6 + 10	(34) 1 + 4	(35) 10 + 7
(36) 8 + 5	(37) 0 + 6	(38) 9 + 9	(39) 8 + 6	(40) 9 + 6

30

Add 0-10

Time: ____________

Score: ______/ 40

① 7 + 2	② 6 + 1	③ 4 + 0	④ 3 + 0	⑤ 5 + 9
⑥ 0 + 4	⑦ 6 + 5	⑧ 8 + 10	⑨ 6 + 2	⑩ 8 + 9
⑪ 8 + 5	⑫ 3 + 5	⑬ 6 + 8	⑭ 9 + 10	⑮ 4 + 0
⑯ 0 + 8	⑰ 10 + 10	⑱ 3 + 4	⑲ 2 + 6	⑳ 0 + 4
(21) 0 + 5	(22) 4 + 3	(23) 7 + 9	(24) 6 + 0	(25) 2 + 9
(26) 10 + 4	(27) 6 + 10	(28) 7 + 1	(29) 2 + 5	(30) 4 + 10
(31) 5 + 3	(32) 2 + 0	(33) 7 + 8	(34) 6 + 6	(35) 5 + 7
(36) 1 + 7	(37) 8 + 0	(38) 7 + 4	(39) 0 + 8	(40) 7 + 10

31

Add 0-10

Time: ________

Score: _____/ 40

(1) 9 + 6	(2) 10 + 7	(3) 10 + 6	(4) 8 + 4	(5) 5 + 5
(6) 0 + 6	(7) 10 + 5	(8) 9 + 8	(9) 5 + 4	(10) 9 + 5
(11) 9 + 7	(12) 1 + 4	(13) 1 + 0	(14) 9 + 2	(15) 1 + 3
(16) 4 + 8	(17) 8 + 7	(18) 3 + 6	(19) 9 + 4	(20) 10 + 8
(21) 8 + 8	(22) 6 + 1	(23) 0 + 2	(24) 7 + 6	(25) 4 + 1
(26) 6 + 3	(27) 3 + 3	(28) 4 + 6	(29) 0 + 9	(30) 2 + 3
(31) 6 + 4	(32) 10 + 1	(33) 8 + 6	(34) 1 + 4	(35) 9 + 1
(36) 3 + 8	(37) 4 + 7	(38) 1 + 10	(39) 2 + 1	(40) 3 + 6

32

Add 0-10

Time: __________

Score: _____/ 40

① 4 + 10	② 7 + 5	③ 0 + 1	④ 5 + 2	⑤ 5 + 8
⑥ 0 + 0	⑦ 10 + 0	⑧ 3 + 1	⑨ 0 + 3	⑩ 8 + 3
⑪ 5 + 1	⑫ 5 + 0	⑬ 6 + 9	⑭ 9 + 3	⑮ 10 + 9
⑯ 7 + 2	⑰ 4 + 8	⑱ 1 + 8	⑲ 3 + 7	⑳ 2 + 7
(21) 1 + 9	(22) 7 + 3	(23) 10 + 2	(24) 0 + 10	(25) 7 + 7
(26) 4 + 9	(27) 8 + 4	(28) 3 + 9	(29) 8 + 1	(30) 4 + 5
(31) 7 + 2	(32) 1 + 5	(33) 2 + 2	(34) 5 + 6	(35) 1 + 1
(36) 10 + 0	(37) 5 + 10	(38) 9 + 0	(39) 2 + 10	(40) 2 + 8

33

Add 0-10

Time: ________

Score: _____/ 40

(1) 9 + 9	(2) 1 + 6	(3) 6 + 1	(4) 7 + 0	(5) 8 + 7
(6) 1 + 7	(7) 8 + 4	(8) 6 + 9	(9) 7 + 5	(10) 5 + 4
(11) 1 + 5	(12) 7 + 9	(13) 6 + 8	(14) 9 + 0	(15) 2 + 1
(16) 10 + 5	(17) 5 + 1	(18) 0 + 0	(19) 9 + 6	(20) 4 + 3
(21) 1 + 6	(22) 3 + 4	(23) 6 + 1	(24) 1 + 9	(25) 7 + 2
(26) 0 + 5	(27) 2 + 10	(28) 4 + 3	(29) 6 + 7	(30) 7 + 4
(31) 9 + 5	(32) 2 + 5	(33) 2 + 3	(34) 10 + 2	(35) 8 + 5
(36) 8 + 8	(37) 9 + 9	(38) 3 + 3	(39) 2 + 8	(40) 3 + 2

34

Add 0-10

Time: __________

Score: _____/ 40

(1) 0 + 2	(2) 1 + 4	(3) 0 + 10	(4) 9 + 2	(5) 4 + 5
(6) 7 + 8	(7) 10 + 4	(8) 3 + 10	(9) 7 + 1	(10) 2 + 6
(11) 4 + 0	(12) 6 + 3	(13) 4 + 1	(14) 7 + 3	(15) 1 + 10
(16) 9 + 7	(17) 6 + 10	(18) 3 + 0	(19) 3 + 6	(20) 10 + 3
(21) 10 + 0	(22) 10 + 8	(23) 6 + 5	(24) 9 + 8	(25) 9 + 1
(26) 5 + 3	(27) 5 + 0	(28) 4 + 10	(29) 10 + 4	(30) 8 + 3
(31) 10 + 1	(32) 1 + 4	(33) 1 + 1	(34) 4 + 9	(35) 5 + 8
(36) 6 + 2	(37) 10 + 6	(38) 0 + 8	(39) 2 + 0	(40) 1 + 10

35

Add 0-10

Time: __________

Score: _____/ 40

① 1 + 2	② 0 + 6	③ 6 + 0	④ 3 + 8	⑤ 9 + 3
⑥ 3 + 9	⑦ 0 + 9	⑧ 5 + 5	⑨ 9 + 4	⑩ 7 + 7
⑪ 5 + 2	⑫ 3 + 5	⑬ 3 + 1	⑭ 8 + 10	⑮ 5 + 10
⑯ 6 + 4	⑰ 1 + 0	⑱ 2 + 9	⑲ 9 + 10	⑳ 5 + 9
(21) 10 + 9	(22) 0 + 3	(23) 6 + 6	(24) 8 + 0	(25) 4 + 2
(26) 4 + 6	(27) 5 + 6	(28) 4 + 8	(29) 0 + 1	(30) 2 + 7
(31) 3 + 7	(32) 3 + 4	(33) 2 + 1	(34) 10 + 7	(35) 2 + 4
(36) 7 + 10	(37) 3 + 6	(38) 1 + 3	(39) 0 + 6	(40) 8 + 6

36

Add 0-10

Time: __________

Score: _____/ 40

① 10 + 10	② 8 + 9	③ 5 + 7	④ 4 + 1	⑤ 4 + 7
⑥ 0 + 9	⑦ 0 + 8	⑧ 2 + 5	⑨ 1 + 9	⑩ 10 + 4
⑪ 2 + 7	⑫ 3 + 5	⑬ 2 + 4	⑭ 4 + 0	⑮ 4 + 2
⑯ 4 + 10	⑰ 2 + 4	⑱ 1 + 4	⑲ 9 + 5	⑳ 8 + 8
(21) 5 + 5	(22) 7 + 4	(23) 9 + 0	(24) 8 + 0	(25) 4 + 1
(26) 4 + 2	(27) 6 + 5	(28) 0 + 8	(29) 9 + 9	(30) 5 + 9
(31) 7 + 8	(32) 7 + 2	(33) 9 + 1	(34) 0 + 2	(35) 10 + 5
(36) 8 + 4	(37) 1 + 7	(38) 5 + 8	(39) 5 + 0	(40) 10 + 7

37

Add 0-10

Time: ________

Score: _____/ 40

① 10 + 1	② 5 + 10	③ 7 + 0	④ 9 + 8	⑤ 9 + 3
⑥ 5 + 4	⑦ 4 + 7	⑧ 6 + 9	⑨ 2 + 6	⑩ 7 + 3
⑪ 7 + 5	⑫ 9 + 10	⑬ 7 + 9	⑭ 8 + 0	⑮ 9 + 2
⑯ 4 + 6	⑰ 7 + 1	⑱ 5 + 2	⑲ 10 + 8	⑳ 6 + 2
(21) 8 + 5	(22) 8 + 6	(23) 6 + 6	(24) 10 + 6	(25) 2 + 8
(26) 0 + 5	(27) 0 + 4	(28) 9 + 6	(29) 5 + 6	(30) 8 + 7
(31) 5 + 3	(32) 4 + 0	(33) 7 + 7	(34) 7 + 10	(35) 3 + 0
(36) 10 + 10	(37) 8 + 10	(38) 1 + 3	(39) 8 + 1	(40) 2 + 0

38

Add 0-10

Time: ____________

Score: ______/ 40

(1) 8 + 3	(2) 4 + 5	(3) 0 + 6	(4) 6 + 8	(5) 0 + 0
(6) 6 + 7	(7) 10 + 2	(8) 1 + 10	(9) 6 + 3	(10) 6 + 10
(11) 3 + 9	(12) 8 + 9	(13) 1 + 4	(14) 2 + 10	(15) 1 + 8
(16) 4 + 8	(17) 9 + 7	(18) 10 + 0	(19) 1 + 0	(20) 3 + 6
(21) 4 + 9	(22) 8 + 4	(23) 10 + 1	(24) 3 + 1	(25) 0 + 10
(26) 1 + 5	(27) 3 + 8	(28) 3 + 7	(29) 6 + 3	(30) 8 + 1
(31) 9 + 4	(32) 3 + 3	(33) 5 + 7	(34) 6 + 0	(35) 0 + 1
(36) 2 + 9	(37) 1 + 1	(38) 4 + 8	(39) 2 + 1	(40) 5 + 1

39

Add 0-10

Time: ________

Score: ____/ 40

(1) 4 + 10	(2) 2 + 2	(3) 2 + 10	(4) 0 + 3	(5) 3 + 4
(6) 10 + 9	(7) 3 + 8	(8) 7 + 3	(9) 1 + 1	(10) 10 + 2
(11) 9 + 7	(12) 6 + 5	(13) 2 + 9	(14) 10 + 4	(15) 1 + 2
(16) 2 + 10	(17) 2 + 3	(18) 4 + 9	(19) 8 + 5	(20) 0 + 7
(21) 1 + 3	(22) 8 + 3	(23) 0 + 0	(24) 5 + 10	(25) 9 + 10
(26) 5 + 2	(27) 8 + 9	(28) 8 + 10	(29) 10 + 10	(30) 5 + 3
(31) 10 + 2	(32) 1 + 10	(33) 2 + 1	(34) 10 + 3	(35) 2 + 0
(36) 1 + 2	(37) 8 + 6	(38) 4 + 2	(39) 2 + 4	(40) 7 + 10

40

Add 0-10

Time: __________

Score: _____/ 40

(1) 4 + 3	(2) 3 + 6	(3) 9 + 8	(4) 3 + 4	(5) 2 + 7
(6) 7 + 1	(7) 1 + 9	(8) 6 + 10	(9) 4 + 7	(10) 1 + 6
(11) 9 + 1	(12) 5 + 9	(13) 1 + 5	(14) 6 + 1	(15) 9 + 2
(16) 1 + 6	(17) 9 + 5	(18) 0 + 3	(19) 2 + 6	(20) 7 + 9
(21) 4 + 3	(22) 0 + 6	(23) 5 + 7	(24) 2 + 5	(25) 6 + 1
(26) 3 + 1	(27) 7 + 4	(28) 5 + 8	(29) 0 + 2	(30) 5 + 0
(31) 6 + 0	(32) 8 + 7	(33) 3 + 0	(34) 6 + 7	(35) 3 + 9
(36) 1 + 7	(37) 5 + 4	(38) 5 + 1	(39) 9 + 4	(40) 8 + 4

41

Add 0-15

Time: __________

Score: _____/ 40

(1) 3 + 12	(2) 11 + 7	(3) 9 + 7	(4) 3 + 8	(5) 1 + 15
(6) 0 + 14	(7) 0 + 13	(8) 12 + 5	(9) 0 + 12	(10) 3 + 2
(11) 9 + 6	(12) 12 + 4	(13) 0 + 5	(14) 2 + 12	(15) 12 + 3
(16) 8 + 9	(17) 0 + 1	(18) 9 + 8	(19) 8 + 2	(20) 10 + 6
(21) 6 + 1	(22) 3 + 13	(23) 3 + 15	(24) 7 + 5	(25) 4 + 9
(26) 3 + 11	(27) 9 + 3	(28) 7 + 6	(29) 10 + 9	(30) 4 + 12
(31) 12 + 8	(32) 2 + 11	(33) 10 + 10	(34) 5 + 7	(35) 13 + 0
(36) 11 + 3	(37) 5 + 13	(38) 3 + 3	(39) 1 + 12	(40) 1 + 6

42

Add 0-15

Time: __________

Score: _____/ 40

(1) 14 + 4	(2) 4 + 3	(3) 11 + 8	(4) 14 + 0	(5) 13 + 1
(6) 2 + 9	(7) 4 + 8	(8) 0 + 11	(9) 6 + 6	(10) 5 + 15
(11) 8 + 3	(12) 5 + 2	(13) 2 + 5	(14) 4 + 14	(15) 8 + 4
(16) 12 + 3	(17) 4 + 2	(18) 7 + 0	(19) 6 + 11	(20) 6 + 12
(21) 4 + 8	(22) 8 + 4	(23) 10 + 1	(24) 2 + 4	(25) 2 + 10
(26) 6 + 14	(27) 6 + 5	(28) 5 + 3	(29) 8 + 11	(30) 2 + 15
(31) 3 + 4	(32) 7 + 12	(33) 4 + 10	(34) 0 + 15	(35) 0 + 14
(36) 1 + 9	(37) 1 + 5	(38) 4 + 13	(39) 2 + 1	(40) 5 + 0

43

Add 0-15

Time: ________

Score: _____/ 40

(1) 10 + 7	(2) 8 + 8	(3) 8 + 7	(4) 0 + 4	(5) 3 + 7
(6) 2 + 14	(7) 3 + 6	(8) 11 + 6	(9) 6 + 8	(10) 5 + 12
(11) 7 + 3	(12) 5 + 8	(13) 13 + 0	(14) 9 + 10	(15) 5 + 4
(16) 15 + 2	(17) 7 + 11	(18) 4 + 6	(19) 0 + 8	(20) 14 + 6
(21) 0 + 6	(22) 7 + 9	(23) 2 + 13	(24) 11 + 9	(25) 1 + 3
(26) 13 + 5	(27) 6 + 0	(28) 13 + 6	(29) 7 + 7	(30) 6 + 4
(31) 11 + 2	(32) 2 + 7	(33) 10 + 4	(34) 4 + 7	(35) 4 + 6
(36) 11 + 1	(37) 8 + 12	(38) 12 + 0	(39) 1 + 1	(40) 5 + 9

44

Add 0-15

Time: ________

Score: _____/ 40

(1) 5 + 14	(2) 14 + 3	(3) 4 + 4	(4) 1 + 11	(5) 10 + 5
(6) 2 + 6	(7) 1 + 13	(8) 3 + 0	(9) 3 + 1	(10) 0 + 7
(11) 7 + 13	(12) 9 + 11	(13) 14 + 5	(14) 4 + 4	(15) 3 + 14
(16) 0 + 9	(17) 11 + 5	(18) 9 + 2	(19) 1 + 8	(20) 9 + 1
(21) 15 + 4	(22) 0 + 3	(23) 4 + 4	(24) 7 + 4	(25) 7 + 8
(26) 13 + 7	(27) 3 + 9	(28) 5 + 11	(29) 8 + 0	(30) 7 + 0
(31) 1 + 7	(32) 6 + 13	(33) 0 + 8	(34) 5 + 14	(35) 3 + 5
(36) 6 + 4	(37) 5 + 5	(38) 8 + 4	(39) 15 + 3	(40) 0 + 10

45

Add 0-15

Time: ____________

Score: _____/ 40

(1) 4 + 11	(2) 4 + 5	(3) 13 + 4	(4) 3 + 14	(5) 13 + 3
(6) 9 + 0	(7) 7 + 10	(8) 2 + 8	(9) 9 + 4	(10) 6 + 9
(11) 5 + 1	(12) 5 + 6	(13) 0 + 2	(14) 6 + 10	(15) 9 + 5
(16) 8 + 2	(17) 14 + 2	(18) 8 + 10	(19) 7 + 1	(20) 1 + 14
(21) 15 + 5	(22) 5 + 4	(23) 4 + 15	(24) 9 + 9	(25) 15 + 1
(26) 4 + 1	(27) 6 + 5	(28) 12 + 7	(29) 11 + 4	(30) 10 + 8
(31) 1 + 2	(32) 4 + 12	(33) 6 + 7	(34) 3 + 13	(35) 14 + 3
(36) 3 + 15	(37) 0 + 8	(38) 9 + 6	(39) 15 + 4	(40) 8 + 9

46

Add 0-15

Time: __________

Score: _____/ 40

(1) 5 + 11	(2) 11 + 2	(3) 7 + 0	(4) 3 + 0	(5) 10 + 10
(6) 15 + 5	(7) 14 + 0	(8) 7 + 9	(9) 12 + 0	(10) 8 + 5
(11) 0 + 9	(12) 13 + 1	(13) 6 + 9	(14) 6 + 8	(15) 7 + 7
(16) 7 + 4	(17) 0 + 10	(18) 5 + 10	(19) 5 + 5	(20) 5 + 0
(21) 13 + 2	(22) 1 + 15	(23) 10 + 0	(24) 2 + 13	(25) 0 + 14
(26) 3 + 1	(27) 5 + 6	(28) 13 + 2	(29) 8 + 7	(30) 0 + 2
(31) 13 + 7	(32) 12 + 5	(33) 2 + 9	(34) 5 + 3	(35) 3 + 11
(36) 0 + 5	(37) 6 + 5	(38) 14 + 4	(39) 2 + 2	(40) 7 + 8

47

Add 0-15

Time: ________

Score: _____/ 40

① 2 + 14	② 4 + 12	③ 3 + 10	④ 1 + 7	⑤ 2 + 11
⑥ 1 + 9	⑦ 0 + 6	⑧ 3 + 2	⑨ 7 + 6	⑩ 13 + 4
⑪ 0 + 7	⑫ 12 + 6	⑬ 10 + 2	⑭ 3 + 3	⑮ 14 + 2
⑯ 12 + 4	⑰ 8 + 3	⑱ 9 + 8	⑲ 2 + 5	⑳ 1 + 12
(21) 5 + 7	(22) 15 + 2	(23) 7 + 0	(24) 9 + 5	(25) 9 + 1
(26) 8 + 11	(27) 2 + 7	(28) 13 + 3	(29) 0 + 4	(30) 4 + 0
(31) 11 + 5	(32) 7 + 1	(33) 4 + 11	(34) 15 + 1	(35) 13 + 0
(36) 0 + 4	(37) 9 + 11	(38) 6 + 7	(39) 0 + 13	(40) 2 + 10

48

Add 0-15

Time: ___________

Score: _____/ 40

① 14 + 6	② 5 + 13	③ 5 + 12	④ 3 + 14	⑤ 13 + 5
⑥ 4 + 6	⑦ 8 + 2	⑧ 11 + 9	⑨ 11 + 4	⑩ 9 + 0
⑪ 13 + 6	⑫ 9 + 9	⑬ 6 + 12	⑭ 4 + 9	⑮ 12 + 1
⑯ 11 + 1	⑰ 5 + 2	⑱ 2 + 15	⑲ 10 + 7	⑳ 2 + 6
(21) 7 + 11	(22) 4 + 4	(23) 14 + 2	(24) 15 + 3	(25) 7 + 5
(26) 4 + 4	(27) 9 + 3	(28) 6 + 13	(29) 5 + 15	(30) 7 + 12
(31) 10 + 6	(32) 0 + 1	(33) 8 + 12	(34) 12 + 6	(35) 2 + 8
(36) 0 + 13	(37) 1 + 12	(38) 13 + 0	(39) 6 + 12	(40) 6 + 14

49

Add 0-15

Time: ____________

Score: _____/ 40

(1) 9 + 4	(2) 3 + 5	(3) 1 + 3	(4) 7 + 10	(5) 9 + 7
(6) 2 + 10	(7) 0 + 11	(8) 11 + 6	(9) 0 + 15	(10) 1 + 6
(11) 6 + 1	(12) 7 + 13	(13) 8 + 6	(14) 10 + 4	(15) 6 + 5
(16) 5 + 9	(17) 4 + 7	(18) 8 + 1	(19) 15 + 4	(20) 10 + 1
(21) 1 + 10	(22) 1 + 4	(23) 11 + 3	(24) 1 + 8	(25) 1 + 13
(26) 12 + 8	(27) 8 + 10	(28) 1 + 6	(29) 4 + 8	(30) 0 + 3
(31) 9 + 10	(32) 11 + 8	(33) 7 + 8	(34) 10 + 9	(35) 1 + 1
(36) 10 + 2	(37) 5 + 1	(38) 3 + 12	(39) 9 + 2	(40) 13 + 1

50

Add 0-15

Time: ________

Score: _____/ 40

(1) 8 + 8	(2) 0 + 14	(3) 3 + 8	(4) 0 + 12	(5) 10 + 5
(6) 8 + 0	(7) 12 + 3	(8) 4 + 10	(9) 6 + 10	(10) 1 + 5
(11) 3 + 6	(12) 6 + 3	(13) 1 + 14	(14) 3 + 9	(15) 6 + 13
(16) 12 + 7	(17) 0 + 8	(18) 11 + 7	(19) 3 + 12	(20) 10 + 3
(21) 1 + 11	(22) 6 + 11	(23) 4 + 5	(24) 10 + 6	(25) 3 + 7
(26) 12 + 2	(27) 14 + 1	(28) 9 + 2	(29) 3 + 4	(30) 4 + 4
(31) 15 + 4	(32) 2 + 15	(33) 0 + 10	(34) 1 + 4	(35) 2 + 12
(36) 15 + 3	(37) 14 + 0	(38) 5 + 4	(39) 4 + 3	(40) 13 + 4

51

Add 0-15

Time: __________

Score: _____/ 40

(1) 4 + 5	(2) 1 + 14	(3) 15 + 2	(4) 9 + 0	(5) 12 + 0
(6) 9 + 7	(7) 4 + 15	(8) 10 + 8	(9) 11 + 8	(10) 5 + 10
(11) 6 + 1	(12) 5 + 6	(13) 14 + 1	(14) 11 + 6	(15) 10 + 6
(16) 0 + 12	(17) 1 + 9	(18) 6 + 6	(19) 1 + 6	(20) 13 + 2
(21) 8 + 12	(22) 15 + 1	(23) 6 + 5	(24) 7 + 10	(25) 5 + 0
(26) 6 + 11	(27) 8 + 0	(28) 6 + 10	(29) 5 + 14	(30) 10 + 10
(31) 4 + 1	(32) 2 + 8	(33) 8 + 2	(34) 8 + 8	(35) 6 + 8
(36) 5 + 1	(37) 10 + 1	(38) 0 + 4	(39) 1 + 3	(40) 8 + 10

52

Add 0-15

Time: ______

Score: _____/ 40

(1) 0 + 15	(2) 5 + 13	(3) 1 + 13	(4) 3 + 13	(5) 5 + 3
(6) 8 + 5	(7) 3 + 9	(8) 13 + 6	(9) 10 + 4	(10) 2 + 3
(11) 13 + 5	(12) 8 + 1	(13) 0 + 12	(14) 4 + 12	(15) 7 + 8
(16) 14 + 6	(17) 9 + 5	(18) 10 + 7	(19) 13 + 1	(20) 1 + 13
(21) 7 + 2	(22) 2 + 11	(23) 8 + 4	(24) 13 + 7	(25) 12 + 5
(26) 5 + 9	(27) 3 + 2	(28) 7 + 0	(29) 1 + 11	(30) 3 + 5
(31) 7 + 4	(32) 5 + 11	(33) 8 + 6	(34) 4 + 8	(35) 0 + 14
(36) 11 + 5	(37) 10 + 2	(38) 4 + 9	(39) 11 + 9	(40) 4 + 10

53

Add 0-15

Time: ________

Score: _____/ 40

(1) 4 + 3	(2) 15 + 5	(3) 14 + 2	(4) 3 + 4	(5) 6 + 13
(6) 1 + 2	(7) 2 + 7	(8) 0 + 3	(9) 13 + 3	(10) 1 + 6
(11) 3 + 15	(12) 8 + 11	(13) 3 + 0	(14) 2 + 2	(15) 6 + 2
(16) 5 + 8	(17) 6 + 14	(18) 8 + 7	(19) 4 + 11	(20) 4 + 8
(21) 1 + 1	(22) 9 + 8	(23) 11 + 3	(24) 4 + 10	(25) 7 + 2
(26) 9 + 11	(27) 1 + 5	(28) 0 + 7	(29) 4 + 6	(30) 6 + 1
(31) 14 + 5	(32) 12 + 1	(33) 0 + 11	(34) 7 + 9	(35) 7 + 12
(36) 11 + 7	(37) 7 + 13	(38) 0 + 5	(39) 1 + 7	(40) 0 + 9

54

Add 0-15

Time: ________

Score: _____/ 40

(1) 5 + 7	(2) 1 + 15	(3) 12 + 8	(4) 2 + 10	(5) 12 + 3
(6) 5 + 12	(7) 11 + 2	(8) 9 + 3	(9) 3 + 1	(10) 7 + 11
(11) 3 + 3	(12) 1 + 12	(13) 3 + 14	(14) 14 + 3	(15) 9 + 9
(16) 2 + 6	(17) 12 + 4	(18) 9 + 4	(19) 12 + 7	(20) 6 + 12
(21) 7 + 5	(22) 3 + 11	(23) 4 + 7	(24) 0 + 2	(25) 3 + 8
(26) 11 + 1	(27) 3 + 6	(28) 10 + 9	(29) 13 + 0	(30) 9 + 1
(31) 8 + 9	(32) 7 + 3	(33) 7 + 7	(34) 2 + 12	(35) 9 + 6
(36) 5 + 15	(37) 2 + 13	(38) 6 + 3	(39) 4 + 12	(40) 0 + 1

55

Add 0-15

Time: ________

Score: _____/ 40

(1) 12 + 6 = ___	(2) 5 + 5 = ___	(3) 9 + 10 = ___	(4) 2 + 14 = ___	(5) 11 + 4 = ___
(6) 4 + 13 = ___	(7) 6 + 9 = ___	(8) 2 + 9 = ___	(9) 0 + 6 = ___	(10) 3 + 2 = ___
(11) 2 + 1 = ___	(12) 2 + 5 = ___	(13) 6 + 10 = ___	(14) 7 + 1 = ___	(15) 10 + 3 = ___
(16) 3 + 12 = ___	(17) 7 + 6 = ___	(18) 8 + 3 = ___	(19) 10 + 6 = ___	(20) 3 + 3 = ___
(21) 4 + 11 = ___	(22) 4 + 9 = ___	(23) 9 + 6 = ___	(24) 0 + 2 = ___	(25) 7 + 12 = ___
(26) 1 + 4 = ___	(27) 3 + 10 = ___	(28) 0 + 3 = ___	(29) 4 + 7 = ___	(30) 15 + 5 = ___
(31) 10 + 5 = ___	(32) 5 + 2 = ___	(33) 11 + 5 = ___	(34) 7 + 10 = ___	(35) 1 + 12 = ___
(36) 2 + 15 = ___	(37) 13 + 4 = ___	(38) 2 + 5 = ___	(39) 7 + 5 = ___	(40) 4 + 13 = ___

56

Add 0-15

Time: ________

Score: _____/ 40

(1) 3 + 4	(2) 7 + 0	(3) 9 + 11	(4) 2 + 6	(5) 5 + 11
(6) 2 + 11	(7) 10 + 9	(8) 5 + 15	(9) 4 + 2	(10) 8 + 12
(11) 12 + 2	(12) 6 + 4	(13) 13 + 3	(14) 13 + 1	(15) 12 + 6
(16) 2 + 2	(17) 8 + 4	(18) 13 + 4	(19) 3 + 14	(20) 4 + 3
(21) 7 + 4	(22) 9 + 2	(23) 1 + 13	(24) 1 + 15	(25) 6 + 12
(26) 12 + 1	(27) 15 + 2	(28) 4 + 6	(29) 2 + 8	(30) 6 + 14
(31) 9 + 10	(32) 1 + 2	(33) 12 + 4	(34) 6 + 6	(35) 5 + 13
(36) 8 + 3	(37) 6 + 2	(38) 8 + 8	(39) 15 + 4	(40) 10 + 6

57

Add 0-15

Time: ________

Score: _____/ 40

(1) 0 + 7 = ___	(2) 3 + 13 = ___	(3) 9 + 4 = ___	(4) 8 + 2 = ___	(5) 8 + 9 = ___
(6) 9 + 7 = ___	(7) 12 + 5 = ___	(8) 1 + 5 = ___	(9) 6 + 7 = ___	(10) 5 + 9 = ___
(11) 2 + 1 = ___	(12) 7 + 1 = ___	(13) 14 + 1 = ___	(14) 4 + 3 = ___	(15) 10 + 1 = ___
(16) 7 + 6 = ___	(17) 5 + 1 = ___	(18) 5 + 14 = ___	(19) 5 + 0 = ___	(20) 6 + 8 = ___
(21) 3 + 2 = ___	(22) 0 + 11 = ___	(23) 6 + 5 = ___	(24) 14 + 2 = ___	(25) 3 + 7 = ___
(26) 2 + 3 = ___	(27) 1 + 1 = ___	(28) 2 + 9 = ___	(29) 0 + 1 = ___	(30) 11 + 8 = ___
(31) 6 + 7 = ___	(32) 0 + 12 = ___	(33) 3 + 4 = ___	(34) 11 + 7 = ___	(35) 12 + 8 = ___
(36) 3 + 2 = ___	(37) 10 + 8 = ___	(38) 5 + 8 = ___	(39) 4 + 8 = ___	(40) 8 + 5 = ___

58

Add 0-15

Time: ________

Score: ______/ 40

① 9 + 9	② 3 + 5	③ 1 + 9	④ 12 + 7	⑤ 7 + 6
⑥ 9 + 0	⑦ 5 + 7	⑧ 1 + 3	⑨ 4 + 5	⑩ 11 + 3
⑪ 6 + 11	⑫ 8 + 6	⑬ 14 + 6	⑭ 11 + 1	⑮ 0 + 5
⑯ 8 + 5	⑰ 0 + 14	⑱ 15 + 3	⑲ 6 + 1	⑳ 3 + 9
(21) 9 + 3	(22) 3 + 0	(23) 10 + 2	(24) 1 + 6	(25) 5 + 6
(26) 0 + 4	(27) 13 + 6	(28) 4 + 12	(29) 9 + 5	(30) 5 + 12
(31) 2 + 14	(32) 5 + 4	(33) 2 + 7	(34) 1 + 13	(35) 0 + 15
(36) 7 + 2	(37) 1 + 11	(38) 3 + 8	(39) 2 + 3	(40) 4 + 14

59

Add 0-15

Time: ____________

Score: _____/ 40

(1) 9 + 8	(2) 10 + 3	(3) 12 + 3	(4) 8 + 7	(5) 8 + 10
(6) 1 + 14	(7) 13 + 0	(8) 11 + 9	(9) 7 + 13	(10) 4 + 1
(11) 0 + 13	(12) 0 + 6	(13) 14 + 4	(14) 5 + 3	(15) 7 + 9
(16) 15 + 1	(17) 14 + 3	(18) 14 + 1	(19) 5 + 2	(20) 10 + 10
(21) 13 + 5	(22) 3 + 11	(23) 3 + 10	(24) 7 + 7	(25) 4 + 15
(26) 6 + 3	(27) 8 + 11	(28) 5 + 10	(29) 0 + 9	(30) 10 + 3
(31) 10 + 7	(32) 3 + 6	(33) 10 + 8	(34) 14 + 5	(35) 1 + 10
(36) 11 + 2	(37) 13 + 7	(38) 7 + 11	(39) 5 + 5	(40) 0 + 10

60

Add 0-15

Time: ________

Score: _____/ 40

(1) 14 + 0	(2) 4 + 4	(3) 9 + 1	(4) 1 + 6	(5) 4 + 11
(6) 3 + 1	(7) 6 + 10	(8) 2 + 12	(9) 1 + 7	(10) 9 + 6
(11) 3 + 15	(12) 11 + 6	(13) 7 + 13	(14) 11 + 4	(15) 10 + 6
(16) 13 + 0	(17) 6 + 1	(18) 4 + 13	(19) 5 + 5	(20) 15 + 3
(21) 5 + 8	(22) 4 + 4	(23) 0 + 13	(24) 12 + 2	(25) 8 + 10
(26) 8 + 0	(27) 1 + 15	(28) 13 + 0	(29) 3 + 5	(30) 4 + 4
(31) 12 + 0	(32) 8 + 4	(33) 9 + 10	(34) 5 + 9	(35) 1 + 9
(36) 10 + 8	(37) 6 + 12	(38) 9 + 0	(39) 1 + 6	(40) 8 + 10

61

Add 0-20

Time: ________

Score: _____/ 40

(1) $7 + 6$	(2) $7 + 5$	(3) $12 + 8$	(4) $2 + 12$	(5) $6 + 8$
(6) $13 + 7$	(7) $2 + 8$	(8) $7 + 2$	(9) $8 + 6$	(10) $7 + 11$
(11) $15 + 4$	(12) $11 + 5$	(13) $1 + 16$	(14) $2 + 7$	(15) $5 + 2$
(16) $12 + 6$	(17) $1 + 0$	(18) $3 + 7$	(19) $18 + 1$	(20) $5 + 11$
(21) $3 + 15$	(22) $11 + 2$	(23) $5 + 4$	(24) $12 + 3$	(25) $9 + 4$
(26) $2 + 4$	(27) $9 + 6$	(28) $11 + 4$	(29) $4 + 2$	(30) $3 + 9$
(31) $1 + 14$	(32) $8 + 9$	(33) $4 + 14$	(34) $1 + 18$	(35) $7 + 12$
(36) $14 + 1$	(37) $4 + 5$	(38) $2 + 4$	(39) $2 + 3$	(40) $1 + 5$

62

Add 0-20

Time: ________

Score: _____/ 40

(1) 6 + 10	(2) 1 + 3	(3) 9 + 7	(4) 11 + 6	(5) 7 + 4
(6) 5 + 8	(7) 1 + 6	(8) 10 + 1	(9) 13 + 6	(10) 10 + 6
(11) 2 + 15	(12) 12 + 5	(13) 4 + 13	(14) 5 + 5	(15) 6 + 11
(16) 14 + 5	(17) 6 + 7	(18) 5 + 14	(19) 8 + 1	(20) 12 + 7
(21) 3 + 10	(22) 19 + 1	(23) 4 + 15	(24) 5 + 1	(25) 12 + 0
(26) 10 + 7	(27) 7 + 10	(28) 2 + 11	(29) 10 + 3	(30) 3 + 3
(31) 11 + 0	(32) 3 + 10	(33) 3 + 5	(34) 7 + 6	(35) 13 + 4
(36) 2 + 1	(37) 13 + 5	(38) 2 + 9	(39) 15 + 2	(40) 12 + 4

63

Add 0-20

Time: ________

Score: _____/ 40

① 11 + 9	② 8 + 4	③ 14 + 1	④ 1 + 13	⑤ 11 + 8
⑥ 16 + 3	⑦ 1 + 11	⑧ 8 + 10	⑨ 13 + 0	⑩ 6 + 12
⑪ 4 + 8	⑫ 8 + 4	⑬ 3 + 6	⑭ 4 + 10	⑮ 16 + 4
⑯ 1 + 14	⑰ 9 + 1	⑱ 6 + 0	⑲ 3 + 8	⑳ 5 + 10
(21) 6 + 5	(22) 16 + 0	(23) 15 + 1	(24) 3 + 12	(25) 9 + 3
(26) 5 + 15	(27) 0 + 13	(28) 10 + 8	(29) 1 + 7	(30) 3 + 1
(31) 12 + 3	(32) 10 + 4	(33) 4 + 1	(34) 5 + 13	(35) 9 + 11
(36) 15 + 5	(37) 6 + 6	(38) 6 + 6	(39) 6 + 3	(40) 11 + 1

64

Add 0-20

Time: ________

Score: _____/ 40

(1) 8 + 3	(2) 8 + 8	(3) 4 + 5	(4) 7 + 10	(5) 10 + 8
(6) 9 + 5	(7) 18 + 2	(8) 3 + 13	(9) 15 + 3	(10) 5 + 6
(11) 6 + 5	(12) 4 + 9	(13) 7 + 7	(14) 15 + 0	(15) 4 + 12
(16) 4 + 4	(17) 2 + 18	(18) 9 + 8	(19) 6 + 4	(20) 3 + 2
(21) 4 + 4	(22) 5 + 12	(23) 5 + 3	(24) 5 + 7	(25) 4 + 6
(26) 17 + 3	(27) 11 + 3	(28) 16 + 1	(29) 3 + 17	(30) 6 + 9
(31) 17 + 2	(32) 17 + 1	(33) 13 + 3	(34) 8 + 12	(35) 7 + 8
(36) 18 + 2	(37) 2 + 6	(38) 1 + 16	(39) 6 + 1	(40) 4 + 11

65

Add 0-20

Time: ________

Score: _____/ 40

(1) 2 + 5	(2) 9 + 10	(3) 14 + 6	(4) 1 + 1	(5) 2 + 2
(6) 9 + 9	(7) 1 + 17	(8) 2 + 17	(9) 3 + 11	(10) 7 + 3
(11) 1 + 9	(12) 12 + 1	(13) 1 + 2	(14) 3 + 7	(15) 6 + 14
(16) 10 + 6	(17) 1 + 19	(18) 8 + 2	(19) 3 + 16	(20) 4 + 3
(21) 10 + 10	(22) 2 + 16	(23) 9 + 0	(24) 7 + 13	(25) 7 + 1
(26) 5 + 9	(27) 7 + 9	(28) 10 + 9	(29) 8 + 11	(30) 6 + 2
(31) 1 + 4	(32) 8 + 6	(33) 14 + 2	(34) 3 + 4	(35) 16 + 3
(36) 1 + 15	(37) 5 + 0	(38) 10 + 5	(39) 9 + 2	(40) 11 + 7

66

Add 0-20

Time: ________

Score: _____/ 40

(1) 7 + 7 = ___	(2) 10 + 2 = ___	(3) 5 + 8 = ___	(4) 10 + 5 = ___	(5) 1 + 9 = ___
(6) 9 + 1 = ___	(7) 2 + 10 = ___	(8) 5 + 0 = ___	(9) 19 + 1 = ___	(10) 11 + 1 = ___
(11) 11 + 3 = ___	(12) 4 + 15 = ___	(13) 11 + 2 = ___	(14) 6 + 5 = ___	(15) 16 + 2 = ___
(16) 6 + 11 = ___	(17) 8 + 2 = ___	(18) 17 + 1 = ___	(19) 10 + 10 = ___	(20) 5 + 14 = ___
(21) 16 + 3 = ___	(22) 11 + 4 = ___	(23) 17 + 2 = ___	(24) 1 + 10 = ___	(25) 13 + 1 = ___
(26) 4 + 2 = ___	(27) 8 + 11 = ___	(28) 2 + 8 = ___	(29) 14 + 5 = ___	(30) 2 + 11 = ___
(31) 2 + 14 = ___	(32) 3 + 8 = ___	(33) 14 + 2 = ___	(34) 3 + 10 = ___	(35) 5 + 6 = ___
(36) 4 + 11 = ___	(37) 2 + 15 = ___	(38) 8 + 2 = ___	(39) 10 + 2 = ___	(40) 6 + 0 = ___

67

Add 0-20

Time: __________

Score: _____/ 40

(1) 8 + 1	(2) 9 + 4	(3) 1 + 18	(4) 4 + 14	(5) 18 + 1
(6) 13 + 6	(7) 14 + 1	(8) 1 + 4	(9) 6 + 8	(10) 10 + 7
(11) 12 + 7	(12) 4 + 5	(13) 1 + 18	(14) 18 + 1	(15) 3 + 12
(16) 2 + 16	(17) 7 + 5	(18) 9 + 11	(19) 5 + 9	(20) 3 + 11
(21) 12 + 0	(22) 9 + 2	(23) 1 + 8	(24) 2 + 6	(25) 1 + 13
(26) 9 + 5	(27) 4 + 12	(28) 3 + 9	(29) 6 + 13	(30) 9 + 9
(31) 9 + 8	(32) 8 + 9	(33) 9 + 10	(34) 0 + 12	(35) 16 + 2
(36) 15 + 4	(37) 2 + 4	(38) 3 + 13	(39) 8 + 6	(40) 5 + 6

68

Add 0-20

Time: ________

Score: _____/ 40

(1) $11 + 6$	(2) $4 + 16$	(3) $8 + 8$	(4) $2 + 2$	(5) $7 + 6$
(6) $3 + 14$	(7) $16 + 4$	(8) $4 + 1$	(9) $2 + 5$	(10) $3 + 3$
(11) $7 + 11$	(12) $1 + 3$	(13) $1 + 15$	(14) $5 + 1$	(15) $7 + 1$
(16) $6 + 7$	(17) $15 + 3$	(18) $12 + 8$	(19) $2 + 7$	(20) $3 + 17$
(21) $6 + 13$	(22) $1 + 19$	(23) $5 + 13$	(24) $1 + 8$	(25) $7 + 12$
(26) $13 + 3$	(27) $6 + 9$	(28) $11 + 9$	(29) $8 + 1$	(30) $9 + 0$
(31) $3 + 1$	(32) $5 + 4$	(33) $14 + 3$	(34) $11 + 0$	(35) $13 + 4$
(36) $3 + 15$	(37) $12 + 4$	(38) $10 + 9$	(39) $5 + 12$	(40) $2 + 10$

69

Add 0-20

Time: ____________

Score: _____/ 40

(1) 7 + 3	(2) 13 + 5	(3) 2 + 13	(4) 5 + 3	(5) 9 + 7
(6) 2 + 3	(7) 5 + 10	(8) 12 + 3	(9) 4 + 6	(10) 7 + 8
(11) 6 + 7	(12) 18 + 2	(13) 1 + 12	(14) 2 + 12	(15) 15 + 0
(16) 14 + 2	(17) 1 + 1	(18) 13 + 7	(19) 2 + 18	(20) 4 + 7
(21) 2 + 1	(22) 4 + 9	(23) 7 + 13	(24) 1 + 6	(25) 5 + 15
(26) 1 + 5	(27) 6 + 10	(28) 3 + 12	(29) 6 + 1	(30) 11 + 5
(31) 15 + 2	(32) 10 + 1	(33) 1 + 0	(34) 1 + 11	(35) 7 + 4
(36) 6 + 2	(37) 9 + 3	(38) 1 + 17	(39) 1 + 7	(40) 12 + 1

70

Add 0-20

Time: ________

Score: _____/ 40

(1) 7 + 2	(2) 2 + 8	(3) 3 + 5	(4) 17 + 3	(5) 1 + 2
(6) 8 + 12	(7) 6 + 14	(8) 9 + 6	(9) 2 + 7	(10) 5 + 11
(11) 3 + 2	(12) 11 + 7	(13) 2 + 9	(14) 14 + 6	(15) 3 + 7
(16) 15 + 1	(17) 7 + 2	(18) 12 + 6	(19) 6 + 12	(20) 15 + 5
(21) 16 + 1	(22) 14 + 4	(23) 5 + 7	(24) 5 + 5	(25) 4 + 13
(26) 12 + 5	(27) 11 + 8	(28) 8 + 3	(29) 7 + 10	(30) 4 + 3
(31) 4 + 8	(32) 1 + 16	(33) 13 + 4	(34) 4 + 13	(35) 2 + 17
(36) 7 + 9	(37) 8 + 4	(38) 4 + 14	(39) 10 + 8	(40) 6 + 1

71

Add 0-20

Time: ___________

Score: _____/ 40

(1) 8 + 2	(2) 12 + 8	(3) 5 + 4	(4) 10 + 8	(5) 9 + 3
(6) 11 + 8	(7) 1 + 19	(8) 10 + 9	(9) 4 + 12	(10) 9 + 5
(11) 4 + 10	(12) 4 + 15	(13) 4 + 5	(14) 1 + 2	(15) 12 + 7
(16) 5 + 4	(17) 1 + 1	(18) 9 + 7	(19) 1 + 6	(20) 9 + 4
(21) 7 + 13	(22) 14 + 5	(23) 10 + 2	(24) 14 + 2	(25) 1 + 15
(26) 5 + 15	(27) 2 + 17	(28) 8 + 1	(29) 5 + 12	(30) 4 + 5
(31) 3 + 12	(32) 2 + 12	(33) 7 + 12	(34) 5 + 5	(35) 8 + 6
(36) 1 + 8	(37) 3 + 13	(38) 10 + 6	(39) 9 + 0	(40) 10 + 4

72

Add 0-20

Time: ________

Score: _____/ 40

(1) 5 + 10	(2) 10 + 7	(3) 7 + 2	(4) 8 + 3	(5) 11 + 5
(6) 1 + 9	(7) 15 + 2	(8) 3 + 6	(9) 8 + 8	(10) 15 + 1
(11) 6 + 3	(12) 16 + 3	(13) 13 + 2	(14) 11 + 1	(15) 11 + 2
(16) 1 + 7	(17) 14 + 6	(18) 10 + 0	(19) 3 + 16	(20) 2 + 14
(21) 15 + 5	(22) 2 + 15	(23) 6 + 7	(24) 7 + 10	(25) 13 + 6
(26) 5 + 2	(27) 2 + 13	(28) 7 + 11	(29) 8 + 12	(30) 6 + 13
(31) 11 + 0	(32) 5 + 0	(33) 11 + 3	(34) 16 + 3	(35) 6 + 0
(36) 14 + 4	(37) 6 + 9	(38) 13 + 2	(39) 17 + 3	(40) 10 + 10

73

Add 0-20

Time: ____________

Score: _____/ 40

(1) 15 + 3	(2) 17 + 2	(3) 2 + 1	(4) 5 + 11	(5) 14 + 3
(6) 2 + 2	(7) 2 + 11	(8) 7 + 8	(9) 1 + 14	(10) 6 + 2
(11) 7 + 1	(12) 1 + 4	(13) 15 + 4	(14) 6 + 13	(15) 8 + 7
(16) 11 + 6	(17) 3 + 2	(18) 2 + 6	(19) 1 + 13	(20) 6 + 6
(21) 12 + 4	(22) 4 + 14	(23) 1 + 3	(24) 9 + 11	(25) 1 + 0
(26) 3 + 10	(27) 3 + 5	(28) 2 + 16	(29) 4 + 11	(30) 12 + 0
(31) 4 + 7	(32) 4 + 1	(33) 5 + 8	(34) 12 + 2	(35) 3 + 14
(36) 6 + 14	(37) 13 + 3	(38) 7 + 4	(39) 3 + 1	(40) 15 + 0

74

Add 0-20

Time: ________

Score: _____/ 40

(1) 10 + 3	(2) 3 + 14	(3) 16 + 0	(4) 11 + 4	(5) 7 + 5
(6) 7 + 6	(7) 6 + 5	(8) 5 + 6	(9) 17 + 1	(10) 8 + 11
(11) 5 + 7	(12) 1 + 11	(13) 3 + 15	(14) 5 + 14	(15) 11 + 7
(16) 6 + 7	(17) 8 + 5	(18) 16 + 4	(19) 7 + 3	(20) 3 + 9
(21) 2 + 4	(22) 3 + 8	(23) 3 + 10	(24) 0 + 16	(25) 14 + 5
(26) 4 + 9	(27) 9 + 8	(28) 13 + 7	(29) 11 + 9	(30) 4 + 2
(31) 3 + 11	(32) 6 + 10	(33) 1 + 8	(34) 2 + 5	(35) 8 + 1
(36) 2 + 9	(37) 9 + 10	(38) 3 + 7	(39) 5 + 1	(40) 10 + 3

75

Add 0-20

Time: ____________

Score: _____/ 40

(1) 7 + 7	(2) 5 + 3	(3) 6 + 4	(4) 1 + 17	(5) 12 + 3
(6) 3 + 16	(7) 9 + 6	(8) 12 + 6	(9) 10 + 5	(10) 8 + 4
(11) 19 + 1	(12) 6 + 11	(13) 6 + 1	(14) 4 + 6	(15) 1 + 5
(16) 6 + 4	(17) 16 + 1	(18) 2 + 3	(19) 4 + 3	(20) 10 + 1
(21) 1 + 10	(22) 9 + 9	(23) 3 + 7	(24) 18 + 2	(25) 7 + 10
(26) 5 + 13	(27) 12 + 5	(28) 6 + 3	(29) 5 + 9	(30) 8 + 9
(31) 9 + 1	(32) 3 + 3	(33) 4 + 6	(34) 2 + 18	(35) 7 + 9
(36) 9 + 2	(37) 13 + 5	(38) 3 + 17	(39) 10 + 1	(40) 1 + 16

76

Add 0-20

Time: ________

Score: _____/ 40

(1) 6 + 0	(2) 2 + 11	(3) 4 + 15	(4) 11 + 9	(5) 6 + 1
(6) 4 + 11	(7) 13 + 4	(8) 3 + 10	(9) 5 + 3	(10) 5 + 6
(11) 5 + 8	(12) 1 + 3	(13) 3 + 4	(14) 4 + 3	(15) 4 + 14
(16) 7 + 6	(17) 1 + 6	(18) 2 + 14	(19) 6 + 7	(20) 6 + 8
(21) 8 + 6	(22) 10 + 9	(23) 8 + 2	(24) 10 + 3	(25) 3 + 2
(26) 6 + 13	(27) 3 + 1	(28) 2 + 3	(29) 15 + 5	(30) 3 + 4
(31) 6 + 8	(32) 13 + 6	(33) 9 + 11	(34) 1 + 2	(35) 11 + 4
(36) 15 + 0	(37) 1 + 14	(38) 8 + 5	(39) 1 + 10	(40) 8 + 3

77

Add 0-20

Time: ________

Score: _____/ 40

(1) 10 + 5 =	(2) 12 + 8 =	(3) 2 + 4 =	(4) 5 + 12 =	(5) 11 + 3 =
(6) 6 + 9 =	(7) 12 + 0 =	(8) 13 + 5 =	(9) 12 + 5 =	(10) 1 + 0 =
(11) 1 + 5 =	(12) 1 + 7 =	(13) 5 + 5 =	(14) 5 + 10 =	(15) 1 + 18 =
(16) 9 + 10 =	(17) 13 + 1 =	(18) 2 + 1 =	(19) 14 + 6 =	(20) 3 + 13 =
(21) 19 + 1 =	(22) 15 + 1 =	(23) 2 + 12 =	(24) 11 + 6 =	(25) 14 + 1 =
(26) 6 + 14 =	(27) 2 + 6 =	(28) 3 + 8 =	(29) 4 + 1 =	(30) 11 + 2 =
(31) 0 + 12 =	(32) 10 + 7 =	(33) 12 + 2 =	(34) 6 + 6 =	(35) 18 + 1 =
(36) 3 + 17 =	(37) 5 + 0 =	(38) 5 + 9 =	(39) 13 + 3 =	(40) 1 + 13 =

78

Add 0-20

Time: ________

Score: _____/ 40

(1) 7 + 13	(2) 15 + 2	(3) 5 + 15	(4) 1 + 19	(5) 6 + 10
(6) 6 + 5	(7) 9 + 7	(8) 7 + 12	(9) 2 + 7	(10) 10 + 6
(11) 3 + 11	(12) 5 + 1	(13) 15 + 3	(14) 2 + 9	(15) 4 + 8
(16) 8 + 6	(17) 4 + 10	(18) 5 + 11	(19) 7 + 8	(20) 8 + 5
(21) 9 + 2	(22) 14 + 2	(23) 19 + 1	(24) 8 + 4	(25) 4 + 8
(26) 6 + 2	(27) 5 + 14	(28) 14 + 5	(29) 5 + 13	(30) 11 + 5
(31) 1 + 16	(32) 16 + 1	(33) 7 + 3	(34) 3 + 6	(35) 10 + 10
(36) 5 + 4	(37) 6 + 11	(38) 1 + 6	(39) 9 + 9	(40) 1 + 19

79

Add 0-20

Time: ____________

Score: ______/ 40

(1) 11 + 7	(2) 11 + 8	(3) 2 + 17	(4) 14 + 4	(5) 3 + 14
(6) 13 + 0	(7) 0 + 13	(8) 9 + 1	(9) 4 + 6	(10) 16 + 0
(11) 2 + 2	(12) 7 + 9	(13) 6 + 12	(14) 10 + 4	(15) 10 + 8
(16) 9 + 8	(17) 0 + 16	(18) 17 + 2	(19) 17 + 3	(20) 4 + 7
(21) 8 + 10	(22) 8 + 4	(23) 13 + 7	(24) 7 + 7	(25) 5 + 10
(26) 8 + 11	(27) 8 + 12	(28) 7 + 10	(29) 4 + 9	(30) 7 + 11
(31) 10 + 5	(32) 1 + 9	(33) 9 + 5	(34) 9 + 3	(35) 3 + 16
(36) 11 + 1	(37) 3 + 15	(38) 7 + 1	(39) 4 + 10	(40) 1 + 1

80

Add 0-20

Time: ________

Score: _____/ 40

(1) 9 + 4	(2) 9 + 0	(3) 2 + 16	(4) 7 + 2	(5) 3 + 5
(6) 5 + 2	(7) 9 + 6	(8) 8 + 9	(9) 11 + 0	(10) 6 + 4
(11) 3 + 12	(12) 2 + 8	(13) 1 + 13	(14) 16 + 1	(15) 16 + 4
(16) 4 + 6	(17) 14 + 3	(18) 1 + 11	(19) 3 + 3	(20) 4 + 2
(21) 16 + 3	(22) 4 + 16	(23) 3 + 16	(24) 6 + 4	(25) 2 + 7
(26) 1 + 17	(27) 1 + 4	(28) 17 + 1	(29) 12 + 6	(30) 2 + 15
(31) 12 + 7	(32) 12 + 1	(33) 3 + 9	(34) 16 + 3	(35) 1 + 12
(36) 15 + 4	(37) 1 + 15	(38) 8 + 8	(39) 12 + 1	(40) 8 + 2

81

Add 0-20

Time: ___________

Score: _____/ 40

① 11 + 7	② 5 + 4	③ 5 + 6	④ 5 + 0	⑤ 17 + 1
⑥ 9 + 5	⑦ 4 + 4	⑧ 1 + 8	⑨ 1 + 3	⑩ 13 + 3
⑪ 6 + 14	⑫ 3 + 11	⑬ 6 + 6	⑭ 6 + 3	⑮ 3 + 3
⑯ 9 + 4	⑰ 4 + 2	⑱ 12 + 8	⑲ 2 + 10	⑳ 1 + 16
(21) 10 + 2	(22) 7 + 13	(23) 2 + 2	(24) 3 + 14	(25) 3 + 10
(26) 7 + 6	(27) 2 + 1	(28) 6 + 10	(29) 6 + 13	(30) 15 + 0
(31) 10 + 5	(32) 3 + 15	(33) 2 + 17	(34) 16 + 3	(35) 8 + 1
(36) 2 + 4	(37) 8 + 8	(38) 10 + 6	(39) 8 + 12	(40) 16 + 1

82

Add 0-20

Time: ____________

Score: _____/ 40

(1) 3 + 17	(2) 13 + 5	(3) 6 + 12	(4) 8 + 9	(5) 11 + 4
(6) 3 + 2	(7) 2 + 8	(8) 2 + 12	(9) 9 + 6	(10) 2 + 7
(11) 5 + 13	(12) 9 + 3	(13) 5 + 9	(14) 3 + 8	(15) 7 + 2
(16) 6 + 1	(17) 3 + 9	(18) 8 + 6	(19) 13 + 7	(20) 12 + 6
(21) 4 + 1	(22) 2 + 7	(23) 3 + 13	(24) 7 + 9	(25) 2 + 16
(26) 3 + 1	(27) 6 + 12	(28) 2 + 6	(29) 8 + 10	(30) 6 + 11
(31) 16 + 4	(32) 5 + 12	(33) 16 + 2	(34) 7 + 7	(35) 5 + 7
(36) 5 + 3	(37) 4 + 3	(38) 2 + 13	(39) 3 + 4	(40) 1 + 9

83

Add 0-20

Time: __________

Score: _____/ 40

(1) 8 + 4	(2) 2 + 5	(3) 12 + 7	(4) 14 + 1	(5) 1 + 17
(6) 17 + 3	(7) 6 + 8	(8) 8 + 6	(9) 7 + 12	(10) 10 + 9
(11) 7 + 1	(12) 4 + 11	(13) 15 + 1	(14) 10 + 8	(15) 9 + 9
(16) 15 + 3	(17) 1 + 6	(18) 14 + 5	(19) 3 + 12	(20) 5 + 10
(21) 11 + 6	(22) 8 + 2	(23) 2 + 9	(24) 4 + 10	(25) 9 + 11
(26) 7 + 4	(27) 12 + 3	(28) 10 + 5	(29) 1 + 14	(30) 12 + 1
(31) 5 + 15	(32) 13 + 0	(33) 1 + 11	(34) 11 + 3	(35) 9 + 10
(36) 5 + 2	(37) 10 + 7	(38) 2 + 8	(39) 7 + 8	(40) 6 + 0

84

Add 0-20

Time: __________

Score: _____/ 40

(1) 9 + 1	(2) 1 + 1	(3) 12 + 2	(4) 4 + 13	(5) 16 + 1
(6) 4 + 12	(7) 13 + 4	(8) 2 + 18	(9) 6 + 7	(10) 6 + 2
(11) 1 + 4	(12) 12 + 5	(13) 1 + 18	(14) 7 + 5	(15) 7 + 3
(16) 3 + 7	(17) 4 + 5	(18) 8 + 5	(19) 15 + 2	(20) 1 + 5
(21) 2 + 15	(22) 11 + 1	(23) 9 + 2	(24) 1 + 14	(25) 14 + 1
(26) 3 + 5	(27) 18 + 1	(28) 11 + 5	(29) 2 + 3	(30) 4 + 15
(31) 11 + 2	(32) 5 + 11	(33) 15 + 4	(34) 4 + 9	(35) 7 + 11
(36) 4 + 13	(37) 14 + 6	(38) 11 + 0	(39) 10 + 1	(40) 7 + 3

85

Add 0-20

Time: __________

Score: _____/ 40

(1) 18 + 2	(2) 7 + 10	(3) 14 + 2	(4) 6 + 9	(5) 12 + 0
(6) 7 + 2	(7) 15 + 5	(8) 9 + 0	(9) 14 + 4	(10) 10 + 7
(11) 4 + 8	(12) 5 + 1	(13) 1 + 15	(14) 1 + 0	(15) 10 + 4
(16) 11 + 9	(17) 8 + 4	(18) 1 + 2	(19) 10 + 10	(20) 13 + 4
(21) 1 + 12	(22) 17 + 2	(23) 4 + 14	(24) 9 + 7	(25) 19 + 1
(26) 13 + 6	(27) 1 + 19	(28) 2 + 14	(29) 8 + 11	(30) 13 + 1
(31) 1 + 7	(32) 12 + 3	(33) 11 + 8	(34) 2 + 11	(35) 2 + 4
(36) 4 + 8	(37) 6 + 5	(38) 9 + 8	(39) 8 + 3	(40) 5 + 5

86

Add 0-20

Time: ____________

Score: _____/ 40

(1) 3 + 16	(2) 5 + 6	(3) 16 + 1	(4) 10 + 0	(5) 13 + 2
(6) 15 + 5	(7) 5 + 7	(8) 2 + 13	(9) 5 + 1	(10) 14 + 2
(11) 1 + 16	(12) 5 + 14	(13) 13 + 2	(14) 6 + 8	(15) 10 + 2
(16) 11 + 7	(17) 2 + 11	(18) 8 + 6	(19) 5 + 5	(20) 8 + 1
(21) 15 + 1	(22) 13 + 5	(23) 5 + 3	(24) 16 + 3	(25) 8 + 9
(26) 6 + 0	(27) 3 + 3	(28) 7 + 12	(29) 9 + 8	(30) 10 + 4
(31) 2 + 9	(32) 3 + 15	(33) 6 + 14	(34) 9 + 10	(35) 3 + 8
(36) 4 + 11	(37) 8 + 12	(38) 1 + 11	(39) 14 + 5	(40) 7 + 5

87

Add 0-20

Time: ____________

Score: _____/ 40

(1) 4 + 5	(2) 8 + 7	(3) 7 + 5	(4) 11 + 5	(5) 8 + 2
(6) 2 + 8	(7) 8 + 10	(8) 10 + 10	(9) 11 + 9	(10) 6 + 11
(11) 8 + 6	(12) 2 + 5	(13) 5 + 2	(14) 2 + 2	(15) 11 + 3
(16) 3 + 11	(17) 2 + 10	(18) 7 + 9	(19) 9 + 4	(20) 12 + 4
(21) 17 + 1	(22) 7 + 1	(23) 1 + 13	(24) 9 + 2	(25) 2 + 6
(26) 13 + 7	(27) 6 + 2	(28) 7 + 8	(29) 12 + 3	(30) 2 + 3
(31) 8 + 11	(32) 5 + 9	(33) 10 + 1	(34) 11 + 6	(35) 8 + 7
(36) 1 + 6	(37) 10 + 6	(38) 13 + 1	(39) 17 + 2	(40) 13 + 3

88

Add 0-20

Time: ________

Score: _____/ 40

(1) 6 + 1	(2) 5 + 8	(3) 11 + 2	(4) 9 + 3	(5) 3 + 1
(6) 3 + 9	(7) 12 + 1	(8) 7 + 11	(9) 1 + 4	(10) 4 + 16
(11) 7 + 7	(12) 7 + 3	(13) 16 + 4	(14) 4 + 1	(15) 12 + 2
(16) 9 + 7	(17) 8 + 5	(18) 12 + 8	(19) 2 + 17	(20) 8 + 3
(21) 16 + 2	(22) 3 + 6	(23) 5 + 15	(24) 5 + 12	(25) 2 + 1
(26) 5 + 11	(27) 4 + 14	(28) 4 + 7	(29) 7 + 2	(30) 15 + 4
(31) 1 + 12	(32) 14 + 4	(33) 12 + 1	(34) 3 + 17	(35) 3 + 5
(36) 1 + 3	(37) 11 + 1	(38) 19 + 1	(39) 14 + 5	(40) 1 + 0

89

Add 0-20

Time: ____________

Score: _____/ 40

(1) 6 + 5	(2) 13 + 6	(3) 9 + 1	(4) 1 + 15	(5) 7 + 6
(6) 5 + 6	(7) 8 + 8	(8) 3 + 7	(9) 7 + 10	(10) 15 + 3
(11) 6 + 6	(12) 9 + 6	(13) 12 + 7	(14) 1 + 1	(15) 6 + 13
(16) 10 + 3	(17) 1 + 7	(18) 3 + 10	(19) 7 + 13	(20) 10 + 3
(21) 3 + 14	(22) 17 + 3	(23) 2 + 4	(24) 4 + 4	(25) 12 + 5
(26) 11 + 4	(27) 8 + 5	(28) 16 + 3	(29) 14 + 3	(30) 6 + 7
(31) 4 + 15	(32) 5 + 0	(33) 6 + 10	(34) 3 + 10	(35) 4 + 12
(36) 4 + 2	(37) 9 + 0	(38) 6 + 3	(39) 11 + 8	(40) 4 + 6

90

Add 0-20

Time: ________

Score: _____/ 40

(1) 2 + 15	(2) 15 + 2	(3) 3 + 13	(4) 2 + 7	(5) 13 + 0
(6) 0 + 13	(7) 16 + 4	(8) 7 + 4	(9) 6 + 12	(10) 9 + 5
(11) 5 + 13	(12) 2 + 14	(13) 14 + 1	(14) 18 + 1	(15) 13 + 0
(16) 3 + 2	(17) 1 + 14	(18) 14 + 3	(19) 5 + 10	(20) 14 + 1
(21) 4 + 10	(22) 9 + 11	(23) 7 + 2	(24) 10 + 8	(25) 9 + 9
(26) 6 + 9	(27) 14 + 6	(28) 10 + 9	(29) 1 + 17	(30) 10 + 4
(31) 4 + 9	(32) 5 + 2	(33) 11 + 0	(34) 15 + 0	(35) 10 + 5
(36) 0 + 13	(37) 2 + 7	(38) 1 + 5	(39) 1 + 2	(40) 1 + 9

91

Add 0-20

Time: __________

Score: _____/ 40

(1) 3 + 15	(2) 9 + 10	(3) 1 + 3	(4) 4 + 11	(5) 12 + 6
(6) 3 + 2	(7) 10 + 9	(8) 1 + 18	(9) 9 + 11	(10) 3 + 8
(11) 5 + 15	(12) 17 + 3	(13) 13 + 6	(14) 18 + 1	(15) 2 + 15
(16) 14 + 2	(17) 7 + 2	(18) 8 + 2	(19) 1 + 2	(20) 13 + 4
(21) 9 + 4	(22) 3 + 4	(23) 1 + 18	(24) 3 + 1	(25) 18 + 1
(26) 1 + 9	(27) 3 + 9	(28) 12 + 4	(29) 2 + 7	(30) 7 + 2
(31) 15 + 4	(32) 12 + 2	(33) 14 + 3	(34) 4 + 9	(35) 12 + 7
(36) 5 + 4	(37) 7 + 12	(38) 6 + 13	(39) 9 + 2	(40) 1 + 14

92

Add 0-20

Time: ________

Score: _____/ 40

(1) 7 + 9	(2) 2 + 13	(3) 3 + 3	(4) 7 + 3	(5) 6 + 7
(6) 7 + 7	(7) 19 + 1	(8) 4 + 2	(9) 11 + 0	(10) 3 + 5
(11) 4 + 14	(12) 11 + 4	(13) 7 + 11	(14) 6 + 2	(15) 9 + 0
(16) 9 + 3	(17) 14 + 5	(18) 9 + 6	(19) 14 + 4	(20) 1 + 11
(21) 13 + 0	(22) 10 + 10	(23) 16 + 3	(24) 11 + 5	(25) 4 + 6
(26) 12 + 0	(27) 8 + 8	(28) 3 + 16	(29) 4 + 15	(30) 5 + 1
(31) 9 + 9	(32) 15 + 5	(33) 3 + 14	(34) 15 + 3	(35) 1 + 1
(36) 2 + 4	(37) 4 + 1	(38) 10 + 7	(39) 17 + 1	(40) 6 + 9

93

Add 0-20

Time: ________

Score: _____/ 40

(1) 8 + 9	(2) 5 + 2	(3) 2 + 5	(4) 2 + 14	(5) 5 + 13
(6) 6 + 1	(7) 9 + 7	(8) 5 + 7	(9) 13 + 2	(10) 1 + 0
(11) 2 + 16	(12) 1 + 17	(13) 6 + 10	(14) 3 + 16	(15) 2 + 2
(16) 3 + 6	(17) 13 + 5	(18) 16 + 2	(19) 6 + 3	(20) 14 + 2
(21) 1 + 6	(22) 11 + 7	(23) 6 + 1	(24) 5 + 6	(25) 8 + 11
(26) 16 + 1	(27) 3 + 13	(28) 5 + 14	(29) 11 + 2	(30) 15 + 2
(31) 11 + 3	(32) 10 + 8	(33) 5 + 12	(34) 2 + 16	(35) 2 + 10
(36) 8 + 12	(37) 16 + 2	(38) 9 + 1	(39) 2 + 11	(40) 2 + 18

94

Add 0-20

Time: __________

Score: _____/ 40

1. 12 + 3 = ___
2. 1 + 5 = ___
3. 5 + 10 = ___
4. 6 + 11 = ___
5. 2 + 9 = ___
6. 6 + 13 = ___
7. 3 + 17 = ___
8. 2 + 7 = ___
9. 11 + 1 = ___
10. 16 + 4 = ___
11. 12 + 5 = ___
12. 1 + 16 = ___
13. 7 + 5 = ___
14. 6 + 4 = ___
15. 12 + 8 = ___
16. 6 + 5 = ___
17. 2 + 17 = ___
18. 2 + 6 = ___
19. 7 + 2 = ___
20. 15 + 0 = ___
21. 5 + 7 = ___
22. 14 + 6 = ___
23. 7 + 1 = ___
24. 2 + 8 = ___
25. 11 + 6 = ___
26. 3 + 7 = ___
27. 18 + 2 = ___
28. 12 + 1 = ___
29. 5 + 0 = ___
30. 1 + 7 = ___
31. 1 + 8 = ___
32. 16 + 1 = ___
33. 1 + 4 = ___
34. 8 + 7 = ___
35. 1 + 15 = ___
36. 7 + 3 = ___
37. 4 + 6 = ___
38. 11 + 8 = ___
39. 2 + 18 = ___
40. 4 + 16 = ___

95

Add 0-20

Time: ________

Score: _____/ 40

(1) 5 + 5	(2) 10 + 3	(3) 4 + 3	(4) 13 + 7	(5) 6 + 14
(6) 2 + 3	(7) 14 + 4	(8) 15 + 1	(9) 6 + 0	(10) 8 + 6
(11) 8 + 3	(12) 5 + 11	(13) 8 + 5	(14) 7 + 4	(15) 5 + 9
(16) 14 + 1	(17) 5 + 3	(18) 1 + 13	(19) 9 + 5	(20) 4 + 13
(21) 13 + 3	(22) 1 + 12	(23) 10 + 4	(24) 5 + 8	(25) 11 + 9
(26) 10 + 5	(27) 9 + 8	(28) 3 + 11	(29) 12 + 1	(30) 8 + 5
(31) 17 + 2	(32) 3 + 4	(33) 7 + 13	(34) 10 + 1	(35) 6 + 12
(36) 2 + 1	(37) 3 + 10	(38) 7 + 10	(39) 2 + 12	(40) 5 + 10

96

Add 0-20

Time: ________

Score: _____/ 40

(1) 7 + 8 =	(2) 7 + 10 =	(3) 9 + 5 =	(4) 2 + 8 =	(5) 2 + 13 =
(6) 1 + 4 =	(7) 8 + 3 =	(8) 8 + 2 =	(9) 7 + 4 =	(10) 9 + 10 =
(11) 14 + 4 =	(12) 6 + 4 =	(13) 4 + 5 =	(14) 12 + 3 =	(15) 5 + 3 =
(16) 12 + 0 =	(17) 1 + 8 =	(18) 2 + 18 =	(19) 10 + 10 =	(20) 3 + 9 =
(21) 10 + 0 =	(22) 8 + 11 =	(23) 13 + 2 =	(24) 4 + 6 =	(25) 3 + 12 =
(26) 3 + 7 =	(27) 3 + 13 =	(28) 8 + 7 =	(29) 2 + 9 =	(30) 18 + 2 =
(31) 16 + 0 =	(32) 0 + 16 =	(33) 2 + 17 =	(34) 4 + 7 =	(35) 1 + 11 =
(36) 11 + 1 =	(37) 8 + 1 =	(38) 5 + 15 =	(39) 0 + 10 =	(40) 3 + 15 =

97

Add 0-20

Time: ________

Score: _____/ 40

(1) 3 + 8	(2) 9 + 6	(3) 1 + 15	(4) 3 + 10	(5) 12 + 1
(6) 15 + 5	(7) 1 + 3	(8) 3 + 1	(9) 2 + 7	(10) 11 + 0
(11) 12 + 2	(12) 14 + 1	(13) 1 + 19	(14) 10 + 3	(15) 7 + 1
(16) 6 + 11	(17) 9 + 8	(18) 7 + 2	(19) 3 + 10	(20) 11 + 6
(21) 4 + 15	(22) 17 + 2	(23) 8 + 5	(24) 3 + 11	(25) 5 + 8
(26) 3 + 3	(27) 13 + 0	(28) 17 + 1	(29) 9 + 2	(30) 1 + 13
(31) 7 + 7	(32) 7 + 13	(33) 15 + 1	(34) 2 + 3	(35) 12 + 8
(36) 6 + 6	(37) 0 + 13	(38) 6 + 9	(39) 15 + 4	(40) 1 + 5

98

Add 0-20

Time: __________

Score: _____/ 40

(1) 2 + 14	(2) 4 + 4	(3) 15 + 3	(4) 2 + 1	(5) 10 + 3
(6) 5 + 13	(7) 5 + 12	(8) 4 + 9	(9) 12 + 7	(10) 14 + 6
(11) 4 + 6	(12) 6 + 2	(13) 3 + 6	(14) 11 + 7	(15) 15 + 0
(16) 9 + 7	(17) 15 + 2	(18) 14 + 5	(19) 4 + 2	(20) 6 + 14
(21) 8 + 7	(22) 6 + 7	(23) 9 + 4	(24) 7 + 6	(25) 7 + 9
(26) 6 + 10	(27) 6 + 7	(28) 5 + 10	(29) 1 + 17	(30) 9 + 9
(31) 8 + 8	(32) 6 + 3	(33) 1 + 12	(34) 6 + 8	(35) 3 + 10
(36) 10 + 5	(37) 13 + 6	(38) 3 + 5	(39) 14 + 2	(40) 5 + 14

99

Add 0-20

Time: ______

Score: ____/ 40

① 6 + 0	② 13 + 4	③ 5 + 7	④ 4 + 1	⑤ 4 + 13
⑥ 7 + 3	⑦ 2 + 6	⑧ 4 + 16	⑨ 16 + 4	⑩ 1 + 9
⑪ 5 + 1	⑫ 2 + 11	⑬ 6 + 5	⑭ 14 + 3	⑮ 4 + 12
⑯ 3 + 14	⑰ 4 + 10	⑱ 11 + 4	⑲ 7 + 5	⑳ 5 + 5
(21) 1 + 0	(22) 9 + 3	(23) 11 + 3	(24) 13 + 3	(25) 5 + 6
(26) 6 + 5	(27) 4 + 7	(28) 5 + 0	(29) 13 + 4	(30) 13 + 7
(31) 10 + 4	(32) 9 + 0	(33) 0 + 6	(34) 7 + 12	(35) 10 + 8
(36) 10 + 2	(37) 5 + 14	(38) 3 + 17	(39) 6 + 1	(40) 8 + 10

100

Add 0-20

Time: ________

Score: _____/ 40

(1) 13 + 1	(2) 10 + 5	(3) 7 + 11	(4) 9 + 1	(5) 4 + 8
(6) 10 + 9	(7) 5 + 9	(8) 9 + 11	(9) 8 + 12	(10) 3 + 2
(11) 8 + 1	(12) 4 + 11	(13) 7 + 6	(14) 17 + 3	(15) 10 + 6
(16) 11 + 9	(17) 12 + 5	(18) 2 + 15	(19) 1 + 1	(20) 16 + 1
(21) 3 + 14	(22) 1 + 6	(23) 5 + 4	(24) 8 + 9	(25) 1 + 7
(26) 11 + 2	(27) 1 + 14	(28) 10 + 1	(29) 11 + 8	(30) 19 + 1
(31) 6 + 12	(32) 8 + 10	(33) 1 + 2	(34) 5 + 11	(35) 2 + 5
(36) 11 + 5	(37) 6 + 10	(38) 10 + 7	(39) 5 + 2	(40) 13 + 5

Answer Key # 1

1. 9	2. 4	3. 3	4. 3	5. 6
6. 6	7. 5	8. 3	9. 8	10. 7
11. 10	12. 5	13. 5	14. 3	15. 2
16. 5	17. 6	18. 8	19. 2	20. 7
21. 2	22. 7	23. 4	24. 1	25. 6
26. 1	27. 6	28. 8	29. 5	30. 0
31. 7	32. 5	33. 3	34. 4	35. 3
36. 9	37. 6	38. 6	39. 4	40. 5

Answer Key # 2

1. 9	2. 3	3. 2	4. 7	5. 8
6. 4	7. 10	8. 4	9. 9	10. 4
11. 5	12. 8	13. 2	14. 5	15. 1
16. 3	17. 7	18. 2	19. 6	20. 5
21. 4	22. 7	23. 4	24. 5	25. 1
26. 6	27. 7	28. 8	29. 6	30. 0
31. 5	32. 4	33. 6	34. 1	35. 4
36. 7	37. 5	38. 0	39. 4	40. 4

Answer Key # 3

1. 6	2. 4	3. 5	4. 9	5. 4
6. 5	7. 4	8. 2	9. 8	10. 7
11. 7	12. 6	13. 9	14. 5	15. 3
16. 2	17. 7	18. 8	19. 1	20. 10
21. 5	22. 3	23. 4	24. 6	25. 6
26. 8	27. 5	28. 2	29. 4	30. 10
31. 5	32. 2	33. 5	34. 5	35. 7
36. 6	37. 9	38. 5	39. 4	40. 7

Answer Key # 4

1. 3	2. 7	3. 8	4. 1	5. 8
6. 2	7. 6	8. 5	9. 6	10. 5
11. 4	12. 0	13. 3	14. 2	15. 8
16. 3	17. 6	18. 1	19. 4	20. 6
21. 4	22. 3	23. 7	24. 9	25. 4
26. 2	27. 6	28. 6	29. 3	30. 3
31. 3	32. 2	33. 7	34. 5	35. 1
36. 1	37. 6	38. 7	39. 8	40. 8

Answer Key # 5

1. 3	2. 6	3. 6	4. 5	5. 7
6. 9	7. 5	8. 5	9. 8	10. 9
11. 5	12. 10	13. 2	14. 5	15. 7
16. 4	17. 3	18. 3	19. 0	20. 4
21. 5	22. 6	23. 2	24. 6	25. 6
26. 7	27. 5	28. 4	29. 3	30. 7
31. 5	32. 2	33. 9	34. 0	35. 3
36. 3	37. 1	38. 3	39. 6	40. 4

Answer Key # 6

1. 10	2. 3	3. 6	4. 8	5. 8
6. 5	7. 7	8. 9	9. 1	10. 5
11. 2	12. 5	13. 4	14. 8	15. 7
16. 4	17. 4	18. 6	19. 5	20. 6
21. 6	22. 4	23. 0	24. 4	25. 6
26. 9	27. 3	28. 4	29. 7	30. 10
31. 8	32. 3	33. 7	34. 2	35. 5
36. 7	37. 4	38. 4	39. 4	40. 9

Answer Key # 7

1. 5	2. 8	3. 1	4. 5	5. 2
6. 2	7. 6	8. 8	9. 5	10. 5
11. 7	12. 1	13. 5	14. 5	15. 5
16. 6	17. 4	18. 10	19. 6	20. 4
21. 1	22. 5	23. 4	24. 3	25. 5
26. 2	27. 7	28. 5	29. 2	30. 4
31. 6	32. 5	33. 8	34. 5	35. 6
36. 8	37. 4	38. 5	39. 2	40. 5

Answer Key # 8

1. 6	2. 3	3. 6	4. 7	5. 6
6. 6	7. 7	8. 2	9. 5	10. 4
11. 7	12. 6	13. 4	14. 7	15. 6
16. 5	17. 6	18. 2	19. 5	20. 4
21. 5	22. 5	23. 2	24. 6	25. 3
26. 5	27. 2	28. 1	29. 5	30. 5
31. 7	32. 3	33. 4	34. 6	35. 8
36. 2	37. 10	38. 6	39. 5	40. 5

Answer Key # 9

1. 6	2. 6	3. 4	4. 8	5. 2
6. 6	7. 10	8. 6	9. 5	10. 1
11. 6	12. 5	13. 2	14. 2	15. 6
16. 7	17. 5	18. 5	19. 8	20. 3
21. 5	22. 5	23. 6	24. 6	25. 4
26. 5	27. 6	28. 7	29. 4	30. 4
31. 5	32. 4	33. 7	34. 2	35. 5
36. 3	37. 6	38. 8	39. 4	40. 5

Answer Key # 10

1. 4	2. 5	3. 4	4. 5	5. 5
6. 5	7. 7	8. 4	9. 6	10. 10
11. 1	12. 5	13. 6	14. 5	15. 6
16. 2	17. 8	18. 2	19. 5	20. 3
21. 2	22. 8	23. 6	24. 5	25. 4
26. 6	27. 2	28. 7	29. 7	30. 6
31. 6	32. 3	33. 4	34. 5	35. 5
36. 6	37. 2	38. 2	39. 5	40. 7

Answer Key # 11

1. 8	2. 5	3. 6	4. 1	5. 5
6. 4	7. 6	8. 5	9. 5	10. 8
11. 6	12. 6	13. 3	14. 6	15. 3
16. 6	17. 6	18. 2	19. 4	20. 5
21. 5	22. 4	23. 7	24. 10	25. 6
26. 4	27. 5	28. 2	29. 5	30. 7
31. 5	32. 4	33. 2	34. 6	35. 8
36. 6	37. 5	38. 5	39. 5	40. 6

Answer Key # 12

1. 5	2. 8	3. 7	4. 4	5. 3
6. 6	7. 4	8. 5	9. 2	10. 4
11. 6	12. 6	13. 7	14. 6	15. 2
16. 3	17. 4	18. 5	19. 5	20. 4
21. 5	22. 5	23. 7	24. 1	25. 2
26. 6	27. 10	28. 5	29. 5	30. 4
31. 5	32. 6	33. 2	34. 5	35. 5
36. 8	37. 7	38. 3	39. 6	40. 2

Answer Key # 13

1. 6	2. 3	3. 5	4. 5	5. 6
6. 1	7. 5	8. 10	9. 5	10. 2
11. 7	12. 4	13. 4	14. 4	15. 7
16. 8	17. 5	18. 6	19. 6	20. 6
21. 5	22. 2	23. 4	24. 6	25. 3
26. 5	27. 4	28. 5	29. 5	30. 6
31. 7	32. 2	33. 5	34. 5	35. 4
36. 6	37. 4	38. 2	39. 2	40. 7

Answer Key # 14

1. 3	2. 7	3. 4	4. 5	5. 6
6. 2	7. 5	8. 6	9. 1	10. 4
11. 5	12. 6	13. 8	14. 5	15. 8
16. 6	17. 6	18. 5	19. 6	20. 10
21. 2	22. 4	23. 5	24. 4	25. 6
26. 5	27. 6	28. 6	29. 6	30. 6
31. 7	32. 10	33. 8	34. 1	35. 7
36. 3	37. 7	38. 6	39. 5	40. 4

Answer Key # 15

1. 6	2. 8	3. 5	4. 2	5. 4
6. 2	7. 2	8. 3	9. 5	10. 4
11. 5	12. 6	13. 5	14. 5	15. 5
16. 5	17. 6	18. 6	19. 5	20. 8
21. 5	22. 5	23. 7	24. 8	25. 4
26. 4	27. 1	28. 5	29. 6	30. 2
31. 4	32. 4	33. 5	34. 3	35. 3
36. 6	37. 6	38. 7	39. 6	40. 6

Answer Key # 16

1. 5	2. 10	3. 2	4. 5	5. 4
6. 5	7. 7	8. 5	9. 5	10. 6
11. 2	12. 2	13. 2	14. 3	15. 7
16. 6	17. 5	18. 4	19. 5	20. 7
21. 6	22. 4	23. 8	24. 5	25. 6
26. 4	27. 1	28. 2	29. 2	30. 10
31. 6	32. 5	33. 5	34. 6	35. 2
36. 8	37. 5	38. 5	39. 6	40. 5

Answer Key # 17

1. 6	2. 5	3. 4	4. 5	5. 3
6. 7	7. 4	8. 6	9. 2	10. 6
11. 5	12. 6	13. 4	14. 6	15. 3
16. 2	17. 5	18. 7	19. 5	20. 2
21. 5	22. 5	23. 5	24. 5	25. 4
26. 6	27. 6	28. 7	29. 6	30. 8
31. 4	32. 10	33. 3	34. 1	35. 5
36. 5	37. 6	38. 2	39. 7	40. 4

Answer Key # 18

1. 4	2. 6	3. 8	4. 5	5. 4
6. 4	7. 6	8. 3	9. 1	10. 5
11. 6	12. 5	13. 6	14. 7	15. 5
16. 5	17. 2	18. 6	19. 4	20. 8
21. 6	22. 3	23. 2	24. 2	25. 5
26. 4	27. 7	28. 5	29. 5	30. 5
31. 5	32. 8	33. 4	34. 7	35. 6
36. 6	37. 10	38. 6	39. 5	40. 2

Answer Key # 19

1. 5	2. 4	3. 8	4. 2	5. 5
6. 3	7. 6	8. 6	9. 7	10. 5
11. 6	12. 5	13. 6	14. 5	15. 5
16. 7	17. 6	18. 7	19. 2	20. 4
21. 4	22. 2	23. 5	24. 5	25. 5
26. 1	27. 8	28. 4	29. 6	30. 6
31. 4	32. 2	33. 5	34. 10	35. 3
36. 6	37. 8	38. 6	39. 8	40. 6

Answer Key # 20

1. 10	2. 6	3. 1	4. 5	5. 5
6. 4	7. 2	8. 6	9. 7	10. 5
11. 2	12. 3	13. 4	14. 6	15. 6
16. 5	17. 2	18. 3	19. 5	20. 5
21. 4	22. 5	23. 5	24. 6	25. 5
26. 7	27. 2	28. 6	29. 4	30. 7
31. 4	32. 5	33. 10	34. 2	35. 5
36. 5	37. 5	38. 1	39. 4	40. 8

Answer Key # 21

1. 4	2. 6	3. 4	4. 5	5. 13
6. 8	7. 11	8. 19	9. 16	10. 10
11. 10	12. 15	13. 5	14. 10	15. 4
16. 8	17. 18	18. 11	19. 10	20. 10
21. 15	22. 16	23. 8	24. 10	25. 10
26. 13	27. 10	28. 7	29. 17	30. 5
31. 14	32. 8	33. 10	34. 11	35. 20
36. 5	37. 5	38. 8	39. 12	40. 3

Answer Key # 22

1. 17	2. 1	3. 12	4. 13	5. 5
6. 2	7. 14	8. 12	9. 14	10. 8
11. 8	12. 10	13. 6	14. 6	15. 10
16. 3	17. 9	18. 11	19. 13	20. 10
21. 18	22. 6	23. 16	24. 2	25. 13
26. 7	27. 10	28. 1	29. 11	30. 9
31. 11	32. 9	33. 10	34. 6	35. 6
36. 10	37. 13	38. 12	39. 17	40. 18

Answer Key # 23

1. 12	2. 12	3. 8	4. 17	5. 10
6. 15	7. 0	8. 7	9. 16	10. 3
11. 11	12. 12	13. 9	14. 14	15. 7
16. 15	17. 11	18. 4	19. 7	20. 14
21. 8	22. 19	23. 2	24. 15	25. 12
26. 6	27. 11	28. 9	29. 8	30. 9
31. 14	32. 3	33. 11	34. 15	35. 7
36. 14	37. 13	38. 12	39. 16	40. 8

Answer Key # 24

1. 4	2. 16	3. 10	4. 10	5. 8
6. 9	7. 6	8. 10	9. 8	10. 15
11. 12	12. 8	13. 14	14. 17	15. 15
16. 0	17. 17	18. 13	19. 8	20. 9
21. 11	22. 6	23. 2	24. 12	25. 11
26. 12	27. 8	28. 19	29. 20	30. 3
31. 16	32. 5	33. 8	34. 9	35. 14
36. 16	37. 6	38. 8	39. 17	40. 6

Answer Key # 25

1. 14	2. 13	3. 10	4. 15	5. 14
6. 9	7. 12	8. 6	9. 19	10. 10
11. 11	12. 13	13. 9	14. 11	15. 14
16. 7	17. 11	18. 11	19. 10	20. 7
21. 4	22. 7	23. 11	24. 4	25. 11
26. 2	27. 14	28. 1	29. 7	30. 7
31. 4	32. 2	33. 10	34. 9	35. 7
36. 16	37. 12	38. 12	39. 15	40. 12

Answer Key # 26

1. 18	2. 17	3. 9	4. 13	5. 8
6. 10	7. 8	8. 3	9. 10	10. 13
11. 10	12. 5	13. 11	14. 13	15. 6
16. 18	17. 7	18. 9	19. 5	20. 4
21. 13	22. 8	23. 12	24. 9	25. 15
26. 10	27. 7	28. 8	29. 10	30. 16
31. 11	32. 10	33. 13	34. 1	35. 10
36. 3	37. 10	38. 4	39. 6	40. 8

Answer Key # 27

1. 15	2. 18	3. 7	4. 11	5. 10
6. 14	7. 1	8. 19	9. 11	10. 10
11. 8	12. 19	13. 2	14. 7	15. 13
16. 18	17. 8	18. 16	19. 16	20. 10
21. 15	22. 8	23. 15	24. 12	25. 10
26. 7	27. 12	28. 9	29. 6	30. 2
31. 11	32. 13	33. 17	34. 6	35. 14
36. 14	37. 10	38. 6	39. 12	40. 12

Answer Key # 28

1. 12	2. 7	3. 9	4. 10	5. 4
6. 16	7. 9	8. 18	9. 4	10. 11
11. 11	12. 2	13. 11	14. 9	15. 8
16. 3	17. 13	18. 8	19. 8	20. 10
21. 20	22. 15	23. 12	24. 8	25. 5
26. 3	27. 10	28. 3	29. 11	30. 10
31. 15	32. 8	33. 12	34. 13	35. 1
36. 9	37. 10	38. 10	39. 13	40. 5

Answer Key # 29

1. 3	2. 13	3. 10	4. 9	5. 17
6. 10	7. 6	8. 12	9. 11	10. 14
11. 10	12. 17	13. 9	14. 10	15. 9
16. 12	17. 4	18. 0	19. 11	20. 12
21. 5	22. 8	23. 8	24. 4	25. 11
26. 14	27. 10	28. 9	29. 5	30. 15
31. 16	32. 11	33. 16	34. 5	35. 17
36. 13	37. 6	38. 18	39. 14	40. 15

Answer Key # 30

1. 9	2. 7	3. 4	4. 3	5. 14
6. 4	7. 11	8. 18	9. 8	10. 17
11. 13	12. 8	13. 14	14. 19	15. 4
16. 8	17. 20	18. 7	19. 8	20. 4
21. 5	22. 7	23. 16	24. 6	25. 11
26. 14	27. 16	28. 8	29. 7	30. 14
31. 8	32. 2	33. 15	34. 12	35. 12
36. 8	37. 8	38. 11	39. 8	40. 17

Answer Key # 31

1. 15	2. 17	3. 16	4. 12	5. 10
6. 6	7. 15	8. 17	9. 9	10. 14
11. 16	12. 5	13. 1	14. 11	15. 4
16. 12	17. 15	18. 9	19. 13	20. 18
21. 16	22. 7	23. 2	24. 13	25. 5
26. 9	27. 6	28. 10	29. 9	30. 5
31. 10	32. 11	33. 14	34. 5	35. 10
36. 11	37. 11	38. 11	39. 3	40. 9

Answer Key # 32

1. 14	2. 12	3. 1	4. 7	5. 13
6. 0	7. 10	8. 4	9. 3	10. 11
11. 6	12. 5	13. 15	14. 12	15. 19
16. 9	17. 12	18. 9	19. 10	20. 9
21. 10	22. 10	23. 12	24. 10	25. 14
26. 13	27. 12	28. 12	29. 9	30. 9
31. 9	32. 6	33. 4	34. 11	35. 2
36. 10	37. 15	38. 9	39. 12	40. 10

Answer Key # 33

1. 18	2. 7	3. 7	4. 7	5. 15
6. 8	7. 12	8. 15	9. 12	10. 9
11. 6	12. 16	13. 14	14. 9	15. 3
16. 15	17. 6	18. 0	19. 15	20. 7
21. 7	22. 7	23. 7	24. 10	25. 9
26. 5	27. 12	28. 7	29. 13	30. 11
31. 14	32. 7	33. 5	34. 12	35. 13
36. 16	37. 18	38. 6	39. 10	40. 5

Answer Key # 34

1. 2	2. 5	3. 10	4. 11	5. 9
6. 15	7. 14	8. 13	9. 8	10. 8
11. 4	12. 9	13. 5	14. 10	15. 11
16. 16	17. 16	18. 3	19. 9	20. 13
21. 10	22. 18	23. 11	24. 17	25. 10
26. 8	27. 5	28. 14	29. 14	30. 11
31. 11	32. 5	33. 2	34. 13	35. 13
36. 8	37. 16	38. 8	39. 2	40. 11

Answer Key # 35

1. 3	2. 6	3. 6	4. 11	5. 12
6. 12	7. 9	8. 10	9. 13	10. 14
11. 7	12. 8	13. 4	14. 18	15. 15
16. 10	17. 1	18. 11	19. 19	20. 14
21. 19	22. 3	23. 12	24. 8	25. 6
26. 10	27. 11	28. 12	29. 1	30. 9
31. 10	32. 7	33. 3	34. 17	35. 6
36. 17	37. 9	38. 4	39. 6	40. 14

Answer Key # 36

1. 20	2. 17	3. 12	4. 5	5. 11
6. 9	7. 8	8. 7	9. 10	10. 14
11. 9	12. 8	13. 6	14. 4	15. 6
16. 14	17. 6	18. 5	19. 14	20. 16
21. 10	22. 11	23. 9	24. 8	25. 5
26. 6	27. 11	28. 8	29. 18	30. 14
31. 15	32. 9	33. 10	34. 2	35. 15
36. 12	37. 8	38. 13	39. 5	40. 17

Answer Key # 37

1. 11	2. 15	3. 7	4. 17	5. 12
6. 9	7. 11	8. 15	9. 8	10. 10
11. 12	12. 19	13. 16	14. 8	15. 11
16. 10	17. 8	18. 7	19. 18	20. 8
21. 13	22. 14	23. 12	24. 16	25. 10
26. 5	27. 4	28. 15	29. 11	30. 15
31. 8	32. 4	33. 14	34. 17	35. 3
36. 20	37. 18	38. 4	39. 9	40. 2

Answer Key # 38

1. 11	2. 9	3. 6	4. 14	5. 0
6. 13	7. 12	8. 11	9. 9	10. 16
11. 12	12. 17	13. 5	14. 12	15. 9
16. 12	17. 16	18. 10	19. 1	20. 9
21. 13	22. 12	23. 11	24. 4	25. 10
26. 6	27. 11	28. 10	29. 9	30. 9
31. 13	32. 6	33. 12	34. 6	35. 1
36. 11	37. 2	38. 12	39. 3	40. 6

Answer Key # 39

1. 14	2. 4	3. 12	4. 3	5. 7
6. 19	7. 11	8. 10	9. 2	10. 12
11. 16	12. 11	13. 11	14. 14	15. 3
16. 12	17. 5	18. 13	19. 13	20. 7
21. 4	22. 11	23. 0	24. 15	25. 19
26. 7	27. 17	28. 18	29. 20	30. 8
31. 12	32. 11	33. 3	34. 13	35. 2
36. 3	37. 14	38. 6	39. 6	40. 17

Answer Key # 40

1. 7	2. 9	3. 17	4. 7	5. 9
6. 8	7. 10	8. 16	9. 11	10. 7
11. 10	12. 14	13. 6	14. 7	15. 11
16. 7	17. 14	18. 3	19. 8	20. 16
21. 7	22. 6	23. 12	24. 7	25. 7
26. 4	27. 11	28. 13	29. 2	30. 5
31. 6	32. 15	33. 3	34. 13	35. 12
36. 8	37. 9	38. 6	39. 13	40. 12

Answer Key # 41

1. 15	2. 18	3. 16	4. 11	5. 16
6. 14	7. 13	8. 17	9. 12	10. 5
11. 15	12. 16	13. 5	14. 14	15. 15
16. 17	17. 1	18. 17	19. 10	20. 16
21. 7	22. 16	23. 18	24. 12	25. 13
26. 14	27. 12	28. 13	29. 19	30. 16
31. 20	32. 13	33. 20	34. 12	35. 13
36. 14	37. 18	38. 6	39. 13	40. 7

Answer Key # 42

1. 18	2. 7	3. 19	4. 14	5. 14
6. 11	7. 12	8. 11	9. 12	10. 20
11. 11	12. 7	13. 7	14. 18	15. 12
16. 15	17. 6	18. 7	19. 17	20. 18
21. 12	22. 12	23. 11	24. 6	25. 12
26. 20	27. 11	28. 8	29. 19	30. 17
31. 7	32. 19	33. 14	34. 15	35. 14
36. 10	37. 6	38. 17	39. 3	40. 5

Answer Key # 43

1. 17	2. 16	3. 15	4. 4	5. 10
6. 16	7. 9	8. 17	9. 14	10. 17
11. 10	12. 13	13. 13	14. 19	15. 9
16. 17	17. 18	18. 10	19. 8	20. 20
21. 6	22. 16	23. 15	24. 20	25. 4
26. 18	27. 6	28. 19	29. 14	30. 10
31. 13	32. 9	33. 14	34. 11	35. 10
36. 12	37. 20	38. 12	39. 2	40. 14

Answer Key # 44

1. 19	2. 17	3. 8	4. 12	5. 15
6. 8	7. 14	8. 3	9. 4	10. 7
11. 20	12. 20	13. 19	14. 8	15. 17
16. 9	17. 16	18. 11	19. 9	20. 10
21. 19	22. 3	23. 8	24. 11	25. 15
26. 20	27. 12	28. 16	29. 8	30. 7
31. 8	32. 19	33. 8	34. 19	35. 8
36. 10	37. 10	38. 12	39. 18	40. 10

Answer Key # 45

1. 15	2. 9	3. 17	4. 17	5. 16
6. 9	7. 17	8. 10	9. 13	10. 15
11. 6	12. 11	13. 2	14. 16	15. 14
16. 10	17. 16	18. 18	19. 8	20. 15
21. 20	22. 9	23. 19	24. 18	25. 16
26. 5	27. 11	28. 19	29. 15	30. 18
31. 3	32. 16	33. 13	34. 16	35. 17
36. 18	37. 8	38. 15	39. 19	40. 17

Answer Key # 46

1. 16	2. 13	3. 7	4. 3	5. 20
6. 20	7. 14	8. 16	9. 12	10. 13
11. 9	12. 14	13. 15	14. 14	15. 14
16. 11	17. 10	18. 15	19. 10	20. 5
21. 15	22. 16	23. 10	24. 15	25. 14
26. 4	27. 11	28. 15	29. 15	30. 2
31. 20	32. 17	33. 11	34. 8	35. 14
36. 5	37. 11	38. 18	39. 4	40. 15

Answer Key # 47

1. 16	2. 16	3. 13	4. 8	5. 13
6. 10	7. 6	8. 5	9. 13	10. 17
11. 7	12. 18	13. 12	14. 6	15. 16
16. 16	17. 11	18. 17	19. 7	20. 13
21. 12	22. 17	23. 7	24. 14	25. 10
26. 19	27. 9	28. 16	29. 4	30. 4
31. 16	32. 8	33. 15	34. 16	35. 13
36. 4	37. 20	38. 13	39. 13	40. 12

Answer Key # 48

1. 20	2. 18	3. 17	4. 17	5. 18
6. 10	7. 10	8. 20	9. 15	10. 9
11. 19	12. 18	13. 18	14. 13	15. 13
16. 12	17. 7	18. 17	19. 17	20. 8
21. 18	22. 8	23. 16	24. 18	25. 12
26. 8	27. 12	28. 19	29. 20	30. 19
31. 16	32. 1	33. 20	34. 18	35. 10
36. 13	37. 13	38. 13	39. 18	40. 20

Answer Key # 49

1. 13	2. 8	3. 4	4. 17	5. 16
6. 12	7. 11	8. 17	9. 15	10. 7
11. 7	12. 20	13. 14	14. 14	15. 11
16. 14	17. 11	18. 9	19. 19	20. 11
21. 11	22. 5	23. 14	24. 9	25. 14
26. 20	27. 18	28. 7	29. 12	30. 3
31. 19	32. 19	33. 15	34. 19	35. 2
36. 12	37. 6	38. 15	39. 11	40. 14

Answer Key # 50

1. 16	2. 14	3. 11	4. 12	5. 15
6. 8	7. 15	8. 14	9. 16	10. 6
11. 9	12. 9	13. 15	14. 12	15. 19
16. 19	17. 8	18. 18	19. 15	20. 13
21. 12	22. 17	23. 9	24. 16	25. 10
26. 14	27. 15	28. 11	29. 7	30. 8
31. 19	32. 17	33. 10	34. 5	35. 14
36. 18	37. 14	38. 9	39. 7	40. 17

Answer Key # 51

1. 9	2. 15	3. 17	4. 9	5. 12
6. 16	7. 19	8. 18	9. 19	10. 15
11. 7	12. 11	13. 15	14. 17	15. 16
16. 12	17. 10	18. 12	19. 7	20. 15
21. 20	22. 16	23. 11	24. 17	25. 5
26. 17	27. 8	28. 16	29. 19	30. 20
31. 5	32. 10	33. 10	34. 16	35. 14
36. 6	37. 11	38. 4	39. 4	40. 18

Answer Key # 52

1. 15	2. 18	3. 14	4. 16	5. 8
6. 13	7. 12	8. 19	9. 14	10. 5
11. 18	12. 9	13. 12	14. 16	15. 15
16. 20	17. 14	18. 17	19. 14	20. 14
21. 9	22. 13	23. 12	24. 20	25. 17
26. 14	27. 5	28. 7	29. 12	30. 8
31. 11	32. 16	33. 14	34. 12	35. 14
36. 16	37. 12	38. 13	39. 20	40. 14

Answer Key # 53

1. 7	2. 20	3. 16	4. 7	5. 19
6. 3	7. 9	8. 3	9. 16	10. 7
11. 18	12. 19	13. 3	14. 4	15. 8
16. 13	17. 20	18. 15	19. 15	20. 12
21. 2	22. 17	23. 14	24. 14	25. 9
26. 20	27. 6	28. 7	29. 10	30. 7
31. 19	32. 13	33. 11	34. 16	35. 19
36. 18	37. 20	38. 5	39. 8	40. 9

Answer Key # 54

1. 12	2. 16	3. 20	4. 12	5. 15
6. 17	7. 13	8. 12	9. 4	10. 18
11. 6	12. 13	13. 17	14. 17	15. 18
16. 8	17. 16	18. 13	19. 19	20. 18
21. 12	22. 14	23. 11	24. 2	25. 11
26. 12	27. 9	28. 19	29. 13	30. 10
31. 17	32. 10	33. 14	34. 14	35. 15
36. 20	37. 15	38. 9	39. 16	40. 1

Answer Key # 55

1. 18	2. 10	3. 19	4. 16	5. 15
6. 17	7. 15	8. 11	9. 6	10. 5
11. 3	12. 7	13. 16	14. 8	15. 13
16. 15	17. 13	18. 11	19. 16	20. 6
21. 15	22. 13	23. 15	24. 2	25. 19
26. 5	27. 13	28. 3	29. 11	30. 20
31. 15	32. 7	33. 16	34. 17	35. 13
36. 17	37. 17	38. 7	39. 12	40. 17

Answer Key # 56

1. 7	2. 7	3. 20	4. 8	5. 16
6. 13	7. 19	8. 20	9. 6	10. 20
11. 14	12. 10	13. 16	14. 14	15. 18
16. 4	17. 12	18. 17	19. 17	20. 7
21. 11	22. 11	23. 14	24. 16	25. 18
26. 13	27. 17	28. 10	29. 10	30. 20
31. 19	32. 3	33. 16	34. 12	35. 18
36. 11	37. 8	38. 16	39. 19	40. 16

Answer Key # 57

1. 7	2. 16	3. 13	4. 10	5. 17
6. 16	7. 17	8. 6	9. 13	10. 14
11. 3	12. 8	13. 15	14. 7	15. 11
16. 13	17. 6	18. 19	19. 5	20. 14
21. 5	22. 11	23. 11	24. 16	25. 10
26. 5	27. 2	28. 11	29. 1	30. 19
31. 13	32. 12	33. 7	34. 18	35. 20
36. 5	37. 18	38. 13	39. 12	40. 13

Answer Key # 58

1. 18	2. 8	3. 10	4. 19	5. 13
6. 9	7. 12	8. 4	9. 9	10. 14
11. 17	12. 14	13. 20	14. 12	15. 5
16. 13	17. 14	18. 18	19. 7	20. 12
21. 12	22. 3	23. 12	24. 7	25. 11
26. 4	27. 19	28. 16	29. 14	30. 17
31. 16	32. 9	33. 9	34. 14	35. 15
36. 9	37. 12	38. 11	39. 5	40. 18

Answer Key # 59

1. 17	2. 13	3. 15	4. 15	5. 18
6. 15	7. 13	8. 20	9. 20	10. 5
11. 13	12. 6	13. 18	14. 8	15. 16
16. 16	17. 17	18. 15	19. 7	20. 20
21. 18	22. 14	23. 13	24. 14	25. 19
26. 9	27. 19	28. 15	29. 9	30. 13
31. 17	32. 9	33. 18	34. 19	35. 11
36. 13	37. 20	38. 18	39. 10	40. 10

Answer Key # 60

1. 14	2. 8	3. 10	4. 7	5. 15
6. 4	7. 16	8. 14	9. 8	10. 15
11. 18	12. 17	13. 20	14. 15	15. 16
16. 13	17. 7	18. 17	19. 10	20. 18
21. 13	22. 8	23. 13	24. 14	25. 18
26. 8	27. 16	28. 13	29. 8	30. 8
31. 12	32. 12	33. 19	34. 14	35. 10
36. 18	37. 18	38. 9	39. 7	40. 18

Answer Key # 61

1. 13	2. 12	3. 20	4. 14	5. 14
6. 20	7. 10	8. 9	9. 14	10. 18
11. 19	12. 16	13. 17	14. 9	15. 7
16. 18	17. 1	18. 10	19. 19	20. 16
21. 18	22. 13	23. 9	24. 15	25. 13
26. 6	27. 15	28. 15	29. 6	30. 12
31. 15	32. 17	33. 18	34. 19	35. 19
36. 15	37. 9	38. 6	39. 5	40. 6

Answer Key # 62

1. 16	2. 4	3. 16	4. 17	5. 11
6. 13	7. 7	8. 11	9. 19	10. 16
11. 17	12. 17	13. 17	14. 10	15. 17
16. 19	17. 13	18. 19	19. 9	20. 19
21. 13	22. 20	23. 19	24. 6	25. 12
26. 17	27. 17	28. 13	29. 13	30. 6
31. 11	32. 13	33. 8	34. 13	35. 17
36. 3	37. 18	38. 11	39. 17	40. 16

Answer Key # 63

1. 20	2. 12	3. 15	4. 14	5. 19
6. 19	7. 12	8. 18	9. 13	10. 18
11. 12	12. 12	13. 9	14. 14	15. 20
16. 15	17. 10	18. 6	19. 11	20. 15
21. 11	22. 16	23. 16	24. 15	25. 12
26. 20	27. 13	28. 18	29. 8	30. 4
31. 15	32. 14	33. 5	34. 18	35. 20
36. 20	37. 12	38. 12	39. 9	40. 12

Answer Key # 64

1. 11	2. 16	3. 9	4. 17	5. 18
6. 14	7. 20	8. 16	9. 18	10. 11
11. 11	12. 13	13. 14	14. 15	15. 16
16. 8	17. 20	18. 17	19. 10	20. 5
21. 8	22. 17	23. 8	24. 12	25. 10
26. 20	27. 14	28. 17	29. 20	30. 15
31. 19	32. 18	33. 16	34. 20	35. 15
36. 20	37. 8	38. 17	39. 7	40. 15

Answer Key # 65

1. 7	2. 19	3. 20	4. 2	5. 4
6. 18	7. 18	8. 19	9. 14	10. 10
11. 10	12. 13	13. 3	14. 10	15. 20
16. 16	17. 20	18. 10	19. 19	20. 7
21. 20	22. 18	23. 9	24. 20	25. 8
26. 14	27. 16	28. 19	29. 19	30. 8
31. 5	32. 14	33. 16	34. 7	35. 19
36. 16	37. 5	38. 15	39. 11	40. 18

Answer Key # 66

1. 14	2. 12	3. 13	4. 15	5. 10
6. 10	7. 12	8. 5	9. 20	10. 12
11. 14	12. 19	13. 13	14. 11	15. 18
16. 17	17. 10	18. 18	19. 20	20. 19
21. 19	22. 15	23. 19	24. 11	25. 14
26. 6	27. 19	28. 10	29. 19	30. 13
31. 16	32. 11	33. 16	34. 13	35. 11
36. 15	37. 17	38. 10	39. 12	40. 6

Answer Key # 67

1. 9	2. 13	3. 19	4. 18	5. 19
6. 19	7. 15	8. 5	9. 14	10. 17
11. 19	12. 9	13. 19	14. 19	15. 15
16. 18	17. 12	18. 20	19. 14	20. 14
21. 12	22. 11	23. 9	24. 8	25. 14
26. 14	27. 16	28. 12	29. 19	30. 18
31. 17	32. 17	33. 19	34. 12	35. 18
36. 19	37. 6	38. 16	39. 14	40. 11

Answer Key # 68

1. 17	2. 20	3. 16	4. 4	5. 13
6. 17	7. 20	8. 5	9. 7	10. 6
11. 18	12. 4	13. 16	14. 6	15. 8
16. 13	17. 18	18. 20	19. 9	20. 20
21. 19	22. 20	23. 18	24. 9	25. 19
26. 16	27. 15	28. 20	29. 9	30. 9
31. 4	32. 9	33. 17	34. 11	35. 17
36. 18	37. 16	38. 19	39. 17	40. 12

Answer Key # 69

1. 10	2. 18	3. 15	4. 8	5. 16
6. 5	7. 15	8. 15	9. 10	10. 15
11. 13	12. 20	13. 13	14. 14	15. 15
16. 16	17. 2	18. 20	19. 20	20. 11
21. 3	22. 13	23. 20	24. 7	25. 20
26. 6	27. 16	28. 15	29. 7	30. 16
31. 17	32. 11	33. 1	34. 12	35. 11
36. 8	37. 12	38. 18	39. 8	40. 13

Answer Key # 70

1. 9	2. 10	3. 8	4. 20	5. 3
6. 20	7. 20	8. 15	9. 9	10. 16
11. 5	12. 18	13. 11	14. 20	15. 10
16. 16	17. 9	18. 18	19. 18	20. 20
21. 17	22. 18	23. 12	24. 10	25. 17
26. 17	27. 19	28. 11	29. 17	30. 7
31. 12	32. 17	33. 17	34. 17	35. 19
36. 16	37. 12	38. 18	39. 18	40. 7

Answer Key # 71

1. 10	2. 20	3. 9	4. 18	5. 12
6. 19	7. 20	8. 19	9. 16	10. 14
11. 14	12. 19	13. 9	14. 3	15. 19
16. 9	17. 2	18. 16	19. 7	20. 13
21. 20	22. 19	23. 12	24. 16	25. 16
26. 20	27. 19	28. 9	29. 17	30. 9
31. 15	32. 14	33. 19	34. 10	35. 14
36. 9	37. 16	38. 16	39. 9	40. 14

Answer Key # 72

1. 15	2. 17	3. 9	4. 11	5. 16
6. 10	7. 17	8. 9	9. 16	10. 16
11. 9	12. 19	13. 15	14. 12	15. 13
16. 8	17. 20	18. 10	19. 19	20. 16
21. 20	22. 17	23. 13	24. 17	25. 19
26. 7	27. 15	28. 18	29. 20	30. 19
31. 11	32. 5	33. 14	34. 19	35. 6
36. 18	37. 15	38. 15	39. 20	40. 20

Answer Key # 73

1. 18	2. 19	3. 3	4. 16	5. 17
6. 4	7. 13	8. 15	9. 15	10. 8
11. 8	12. 5	13. 19	14. 19	15. 15
16. 17	17. 5	18. 8	19. 14	20. 12
21. 16	22. 18	23. 4	24. 20	25. 1
26. 13	27. 8	28. 18	29. 15	30. 12
31. 11	32. 5	33. 13	34. 14	35. 17
36. 20	37. 16	38. 11	39. 4	40. 15

Answer Key # 74

1. 13	2. 17	3. 16	4. 15	5. 12
6. 13	7. 11	8. 11	9. 18	10. 19
11. 12	12. 12	13. 18	14. 19	15. 18
16. 13	17. 13	18. 20	19. 10	20. 12
21. 6	22. 11	23. 13	24. 16	25. 19
26. 13	27. 17	28. 20	29. 20	30. 6
31. 14	32. 16	33. 9	34. 7	35. 9
36. 11	37. 19	38. 10	39. 6	40. 13

Answer Key # 75

1. 14	2. 8	3. 10	4. 18	5. 15
6. 19	7. 15	8. 18	9. 15	10. 12
11. 20	12. 17	13. 7	14. 10	15. 6
16. 10	17. 17	18. 5	19. 7	20. 11
21. 11	22. 18	23. 10	24. 20	25. 17
26. 18	27. 17	28. 9	29. 14	30. 17
31. 10	32. 6	33. 10	34. 20	35. 16
36. 11	37. 18	38. 20	39. 11	40. 17

Answer Key # 76

1. 6	2. 13	3. 19	4. 20	5. 7
6. 15	7. 17	8. 13	9. 8	10. 11
11. 13	12. 4	13. 7	14. 7	15. 18
16. 13	17. 7	18. 16	19. 13	20. 14
21. 14	22. 19	23. 10	24. 13	25. 5
26. 19	27. 4	28. 5	29. 20	30. 7
31. 14	32. 19	33. 20	34. 3	35. 15
36. 15	37. 15	38. 13	39. 11	40. 11

Answer Key # 77

1. 15	2. 20	3. 6	4. 17	5. 14
6. 15	7. 12	8. 18	9. 17	10. 1
11. 6	12. 8	13. 10	14. 15	15. 19
16. 19	17. 14	18. 3	19. 20	20. 16
21. 20	22. 16	23. 14	24. 17	25. 15
26. 20	27. 8	28. 11	29. 5	30. 13
31. 12	32. 17	33. 14	34. 12	35. 19
36. 20	37. 5	38. 14	39. 16	40. 14

Answer Key # 78

1. 20	2. 17	3. 20	4. 20	5. 16
6. 11	7. 16	8. 19	9. 9	10. 16
11. 14	12. 6	13. 18	14. 11	15. 12
16. 14	17. 14	18. 16	19. 15	20. 13
21. 11	22. 16	23. 20	24. 12	25. 12
26. 8	27. 19	28. 19	29. 18	30. 16
31. 17	32. 17	33. 10	34. 9	35. 20
36. 9	37. 17	38. 7	39. 18	40. 20

Answer Key # 79

1. 18	2. 19	3. 19	4. 18	5. 17
6. 13	7. 13	8. 10	9. 10	10. 16
11. 4	12. 16	13. 18	14. 14	15. 18
16. 17	17. 16	18. 19	19. 20	20. 11
21. 18	22. 12	23. 20	24. 14	25. 15
26. 19	27. 20	28. 17	29. 13	30. 18
31. 15	32. 10	33. 14	34. 12	35. 19
36. 12	37. 18	38. 8	39. 14	40. 2

Answer Key # 80

1. 13	2. 9	3. 18	4. 9	5. 8
6. 7	7. 15	8. 17	9. 11	10. 10
11. 15	12. 10	13. 14	14. 17	15. 20
16. 10	17. 17	18. 12	19. 6	20. 6
21. 19	22. 20	23. 19	24. 10	25. 9
26. 18	27. 5	28. 18	29. 18	30. 17
31. 19	32. 13	33. 12	34. 19	35. 13
36. 19	37. 16	38. 16	39. 13	40. 10

Answer Key # 81

1. 18	2. 9	3. 11	4. 5	5. 18
6. 14	7. 8	8. 9	9. 4	10. 16
11. 20	12. 14	13. 12	14. 9	15. 6
16. 13	17. 6	18. 20	19. 12	20. 17
21. 12	22. 20	23. 4	24. 17	25. 13
26. 13	27. 3	28. 16	29. 19	30. 15
31. 15	32. 18	33. 19	34. 19	35. 9
36. 6	37. 16	38. 16	39. 20	40. 17

Answer Key # 82

1. 20	2. 18	3. 18	4. 17	5. 15
6. 5	7. 10	8. 14	9. 15	10. 9
11. 18	12. 12	13. 14	14. 11	15. 9
16. 7	17. 12	18. 14	19. 20	20. 18
21. 5	22. 9	23. 16	24. 16	25. 18
26. 4	27. 18	28. 8	29. 18	30. 17
31. 20	32. 17	33. 18	34. 14	35. 12
36. 8	37. 7	38. 15	39. 7	40. 10

Answer Key # 83

1. 12	2. 7	3. 19	4. 15	5. 18
6. 20	7. 14	8. 14	9. 19	10. 19
11. 8	12. 15	13. 16	14. 18	15. 18
16. 18	17. 7	18. 19	19. 15	20. 15
21. 17	22. 10	23. 11	24. 14	25. 20
26. 11	27. 15	28. 15	29. 15	30. 13
31. 20	32. 13	33. 12	34. 14	35. 19
36. 7	37. 17	38. 10	39. 15	40. 6

Answer Key # 84

1. 10	2. 2	3. 14	4. 17	5. 17
6. 16	7. 17	8. 20	9. 13	10. 8
11. 5	12. 17	13. 19	14. 12	15. 10
16. 10	17. 9	18. 13	19. 17	20. 6
21. 17	22. 12	23. 11	24. 15	25. 15
26. 8	27. 19	28. 16	29. 5	30. 19
31. 13	32. 16	33. 19	34. 13	35. 18
36. 17	37. 20	38. 11	39. 11	40. 10

Answer Key # 85

1. 20	2. 17	3. 16	4. 15	5. 12
6. 9	7. 20	8. 9	9. 18	10. 17
11. 12	12. 6	13. 16	14. 1	15. 14
16. 20	17. 12	18. 3	19. 20	20. 17
21. 13	22. 19	23. 18	24. 16	25. 20
26. 19	27. 20	28. 16	29. 19	30. 14
31. 8	32. 15	33. 19	34. 13	35. 6
36. 12	37. 11	38. 17	39. 11	40. 10

Answer Key # 86

1. 19	2. 11	3. 17	4. 10	5. 15
6. 20	7. 12	8. 15	9. 6	10. 16
11. 17	12. 19	13. 15	14. 14	15. 12
16. 18	17. 13	18. 14	19. 10	20. 9
21. 16	22. 18	23. 8	24. 19	25. 17
26. 6	27. 6	28. 19	29. 17	30. 14
31. 11	32. 18	33. 20	34. 19	35. 11
36. 15	37. 20	38. 12	39. 19	40. 12

Answer Key # 87

1. 9	2. 15	3. 12	4. 16	5. 10
6. 10	7. 18	8. 20	9. 20	10. 17
11. 14	12. 7	13. 7	14. 4	15. 14
16. 14	17. 12	18. 16	19. 13	20. 16
21. 18	22. 8	23. 14	24. 11	25. 8
26. 20	27. 8	28. 15	29. 15	30. 5
31. 19	32. 14	33. 11	34. 17	35. 15
36. 7	37. 16	38. 14	39. 19	40. 16

Answer Key # 88

1. 7	2. 13	3. 13	4. 12	5. 4
6. 12	7. 13	8. 18	9. 5	10. 20
11. 14	12. 10	13. 20	14. 5	15. 14
16. 16	17. 13	18. 20	19. 19	20. 11
21. 18	22. 9	23. 20	24. 17	25. 3
26. 16	27. 18	28. 11	29. 9	30. 19
31. 13	32. 18	33. 13	34. 20	35. 8
36. 4	37. 12	38. 20	39. 19	40. 1

Answer Key # 89

1. 11	2. 19	3. 10	4. 16	5. 13
6. 11	7. 16	8. 10	9. 17	10. 18
11. 12	12. 15	13. 19	14. 2	15. 19
16. 13	17. 8	18. 13	19. 20	20. 13
21. 17	22. 20	23. 6	24. 8	25. 17
26. 15	27. 13	28. 19	29. 17	30. 13
31. 19	32. 5	33. 16	34. 13	35. 16
36. 6	37. 9	38. 9	39. 19	40. 10

Answer Key # 90

1. 17	2. 17	3. 16	4. 9	5. 13
6. 13	7. 20	8. 11	9. 18	10. 14
11. 18	12. 16	13. 15	14. 19	15. 13
16. 5	17. 15	18. 17	19. 15	20. 15
21. 14	22. 20	23. 9	24. 18	25. 18
26. 15	27. 20	28. 19	29. 18	30. 14
31. 13	32. 7	33. 11	34. 15	35. 15
36. 13	37. 9	38. 6	39. 3	40. 10

Answer Key # 91

1. 18	2. 19	3. 4	4. 15	5. 18
6. 5	7. 19	8. 19	9. 20	10. 11
11. 20	12. 20	13. 19	14. 19	15. 17
16. 16	17. 9	18. 10	19. 3	20. 17
21. 13	22. 7	23. 19	24. 4	25. 19
26. 10	27. 12	28. 16	29. 9	30. 9
31. 19	32. 14	33. 17	34. 13	35. 19
36. 9	37. 19	38. 19	39. 11	40. 15

Answer Key # 92

1. 16	2. 15	3. 6	4. 10	5. 13
6. 14	7. 20	8. 6	9. 11	10. 8
11. 18	12. 15	13. 18	14. 8	15. 9
16. 12	17. 19	18. 15	19. 18	20. 12
21. 13	22. 20	23. 19	24. 16	25. 10
26. 12	27. 16	28. 19	29. 19	30. 6
31. 18	32. 20	33. 17	34. 18	35. 2
36. 6	37. 5	38. 17	39. 18	40. 15

Answer Key # 93

1. 17	2. 7	3. 7	4. 16	5. 18
6. 7	7. 16	8. 12	9. 15	10. 1
11. 18	12. 18	13. 16	14. 19	15. 4
16. 9	17. 18	18. 18	19. 9	20. 16
21. 7	22. 18	23. 7	24. 11	25. 19
26. 17	27. 16	28. 19	29. 13	30. 17
31. 14	32. 18	33. 17	34. 18	35. 12
36. 20	37. 18	38. 10	39. 13	40. 20

Answer Key # 94

1. 15	2. 6	3. 15	4. 17	5. 11
6. 19	7. 20	8. 9	9. 12	10. 20
11. 17	12. 17	13. 12	14. 10	15. 20
16. 11	17. 19	18. 8	19. 9	20. 15
21. 12	22. 20	23. 8	24. 10	25. 17
26. 10	27. 20	28. 13	29. 5	30. 8
31. 9	32. 17	33. 5	34. 15	35. 16
36. 10	37. 10	38. 19	39. 20	40. 20

Answer Key # 95

1. 10	2. 13	3. 7	4. 20	5. 20
6. 5	7. 18	8. 16	9. 6	10. 14
11. 11	12. 16	13. 13	14. 11	15. 14
16. 15	17. 8	18. 14	19. 14	20. 17
21. 16	22. 13	23. 14	24. 13	25. 20
26. 15	27. 17	28. 14	29. 13	30. 13
31. 19	32. 7	33. 20	34. 11	35. 18
36. 3	37. 13	38. 17	39. 14	40. 15

Answer Key # 96

1. 15	2. 17	3. 14	4. 10	5. 15
6. 5	7. 11	8. 10	9. 11	10. 19
11. 18	12. 10	13. 9	14. 15	15. 8
16. 12	17. 9	18. 20	19. 20	20. 12
21. 10	22. 19	23. 15	24. 10	25. 15
26. 10	27. 16	28. 15	29. 11	30. 20
31. 16	32. 16	33. 19	34. 11	35. 12
36. 12	37. 9	38. 20	39. 10	40. 18

Answer Key # 97

1. 11	2. 15	3. 16	4. 13	5. 13
6. 20	7. 4	8. 4	9. 9	10. 11
11. 14	12. 15	13. 20	14. 13	15. 8
16. 17	17. 17	18. 9	19. 13	20. 17
21. 19	22. 19	23. 13	24. 14	25. 13
26. 6	27. 13	28. 18	29. 11	30. 14
31. 14	32. 20	33. 16	34. 5	35. 20
36. 12	37. 13	38. 15	39. 19	40. 6

Answer Key # 98

1. 16	2. 8	3. 18	4. 3	5. 13
6. 18	7. 17	8. 13	9. 19	10. 20
11. 10	12. 8	13. 9	14. 18	15. 15
16. 16	17. 17	18. 19	19. 6	20. 20
21. 15	22. 13	23. 13	24. 13	25. 16
26. 16	27. 13	28. 15	29. 18	30. 18
31. 16	32. 9	33. 13	34. 14	35. 13
36. 15	37. 19	38. 8	39. 16	40. 19

Answer Key # 99

1. 6	2. 17	3. 12	4. 5	5. 17
6. 10	7. 8	8. 20	9. 20	10. 10
11. 6	12. 13	13. 11	14. 17	15. 16
16. 17	17. 14	18. 15	19. 12	20. 10
21. 1	22. 12	23. 14	24. 16	25. 11
26. 11	27. 11	28. 5	29. 17	30. 20
31. 14	32. 9	33. 6	34. 19	35. 18
36. 12	37. 19	38. 20	39. 7	40. 18

Answer Key # 100

1. 14	2. 15	3. 18	4. 10	5. 12
6. 19	7. 14	8. 20	9. 20	10. 5
11. 9	12. 15	13. 13	14. 20	15. 16
16. 20	17. 17	18. 17	19. 2	20. 17
21. 17	22. 7	23. 9	24. 17	25. 8
26. 13	27. 15	28. 11	29. 19	30. 20
31. 18	32. 18	33. 3	34. 16	35. 7
36. 16	37. 16	38. 17	39. 7	40. 18

Track your progress

Set	Time	Score	Set	Time	Score	Set	Time	Score	Set	Time	Score
1			26			51			76		
2			27			52			77		
3			28			53			78		
4			29			54			79		
5			30			55			80		
6			31			56			81		
7			32			57			82		
8			33			58			83		
9			34			59			84		
10			35			60			85		
11			36			61			86		
12			37			62			87		
13			38			63			88		
14			39			64			89		
15			40			65			90		
16			41			66			91		
17			42			67			92		
18			43			68			93		
19			44			69			94		
20			45			70			95		
21			46			71			96		
22			47			72			97		
23			48			73			98		
24			49			74			99		
25			50			75			100		

Made in United States
North Haven, CT
05 November 2026

21777134R00061